FULLERENES AND RELATIVE MATERIALS - PROPERTIES AND APPLICATIONS

Edited by **Natalia V. Kamanina**

Fullerenes and Relative Materials - Properties and Applications
http://dx.doi.org/10.5772/68106
Edited by Natalia V. Kamanina

Contributors

Ming-Der Su, Ming-Chung Yang, Nora Berrah, Bin Zhang, Kaixiong Gao, Yuanlie Yu, Junyan Zhang, Francisco J Tenorio, David Alejandro Hernández-Velázquez, Jaime Gustavo Rodríguez-Zavala, Ernestina Mora Jiménez, Gregorio Guzmán-Ramírez, Abdelali Rahmani, Hassan Chadli, Fatima Fergani, Konstantin Amsharov, Elif Okutan, Natalia Vladimirovna Kamanina

First published in London, United Kingdom, 2018 by IntechOpen
IntechOpen is the global imprint of INTECHOPEN LIMITED, registered in England and Wales, registration number: 11086078, The Shard, 25th floor, 32 London Bridge Street
London, SE19SG – United Kingdom
Printed in Croatia

British Library Cataloguing-in-Publication Data
A catalogue record for this book is available from the British Library

Additional hard copies can be obtained from orders@intechopen.com

Fullerenes and Relative Materials - Properties and Applications, Edited by Natalia V. Kamanina
p. cm.
Print ISBN 978-1-78923-022-2
Online ISBN 978-1-78923-023-9

For EU product safety concerns:
IN TECH d.o.o., Prolaz Marije Krucifikse Kozulić 3, 51000 Rijeka, Croatia,
info@intechopen.com or visit our website at intechopen.com.

Meet the editor

Prof., Dr. Sci., PhD Natalia Vladimirovna Kamanina was born in Kaliningrad, Russian Federation, on March 19, 1957. She graduated from general school with gold medal (1974) and graduated with honor diploma from Leningrad Polytechnic Institute (1981), St. Petersburg, Russia. She has received a PhD degree (Physics and Mathematics) at Vavilov State Optical Institute, St. Petersburg, Russia (1995), and a Dr. Sci. degree (Physics and Mathematics) at Vavilov State Optical Institute, St. Petersburg, Russia (2001). Now, she is the head of the Lab for "Photophysics of media with nanoobjects" at Vavilov State Optical Institute St. Petersburg, Russia. She has been involved in the collaboration research with many researchers and scientists in the world since 1995 and published in the area of laser-matter interaction and nanotechnology in optics about 200 technical papers and 13 patents. She has been the chief of three international and ten Russian scientific projects. In parallel to her scientific activity, she has been lecturing from 2002. She is a professor at St. Petersburg Electrotechnical University "LETI." Under her leadership, 5 young scientists have defended their PhD theses, and 17 master's students have received their master's degree diploma. Now, her current interest includes nano- and biotechnology in optics.

Contents

Preface

This book entitled *Fullerenes and Relative Materials: Properties and Applications* considers some aspects of the synthesis, structural, vibrational, tribology, optical, and nonlinear optical properties of the fullerenes and relative materials to some area of the applications. It consists of eight chapters including the introductory one. All chapters present the results of the unique properties of the materials studied in the different areas of their applications, including the general optoelectronics, solar energy and gas storage, laser and display, and biomedicine. It is important for the education process and for civil and special device operations. The advanced idea, the special approach, and the information described in the current book will be fruitful for readers to find a sustainable solution in a fundamental study and in the technique. The book can be useful for students, postgraduate students, engineers, researchers, and technical officers of optoelectronic universities and companies.

The editor of this book would like to thank all the authors of the chapters presented and to acknowledge the reviewers and all who have helped prepare this book. The editor would like to express gratitude to Mr. Slobodan Momcilovic, Publishing Process Manager, IntechOpen, Croatia, for his good and continued cooperation.

N. V. Kamanina, Dr.Sc., PhD
Head of the Lab for "Photophysics of media with nanoobjects"
Vavilov State Optical Institute
Saint Petersburg, Russia

Professor at the St.-Petersburg Electrotechnical University ("LETI")
Saint Petersburg, Russia

Chapter 1

Introductory Chapter: Why the Study of Fullerenes is so Important?

Natalia Vladimirovna Kamanina

Additional information is available at the end of the chapter

http://dx.doi.org/10.5772/intechopen.74812

1. Introduction

It is known that the use of the laser technique has been dramatically activated from the 1960's. Due to the reason that the lasers are operated at different spectral range and at different energy density (power), the scientists and engineers have had the task to find the novel materials to protect the human eyes and technical devices from high laser irradiation. Moreover, the modern solar energy, gas storage, and biomedicine elements have searched for the new sensitizers with good advantage as well.

In this aspect, the science history and nature permits to resolve this task [1–7]. From David Jones experiments (1966) and E. Osawa calculations (1970); from E. Colder experiments (1984), and from H.W. Kroto, R.E. Smalley, R.F. Curl consideration (1985), the fullerenes have been discovered and have got their name on 1985. Indeed, it provokes the future investigation. Thus, the carbon nanotubes (CNTs) have been opened in 1991. Now, these nanostructures have an important advantage in the modification of the surface and the body of inorganic and organic materials.

The main reason to use the fullerenes is connected with their unique energy levels and high value of electron affinity energy. Only S_1-T_1 transition has the time close to 1.2 ns, but the higher transitions from the exited singlet or triplet states have the time in the pico- or femtosecond range. Thus, the study of the fullerene and relative systems dynamic in the nano-, pico-, and femtosecond regime is actual, and the investigation of the ionization and fragmentation of fullerenes with different approaches are important. Moreover, the fullerenes have the large value of the electron affinity energy close to 0.65–0.7 eV. That is larger than the one for most dyes and organic molecules *intra*molecular acceptor fragment. It can stimulate the efficient *inter*-molecular charge transfer complex (CTC) formation in the fullerene-doped organic conjugated materials. Regarding the CNTs, it should be necessary to take into account the variety of charge transfer pathways, including those along and across a CNT, between CNTs, inside a multiwall CNT, between organic molecules and CNTs, and between the donor and acceptor moieties of an organic matrix molecule. The data presented in **Figure 1a** and **b** show the possible mechanism of charge carriers moving in the fullerene-, carbon nanotubes, and relative nano-objects-doped organic materials. Moreover, the bio-objects based on DNA can be added to this consideration. Some theoretical and experimental results about the process shown in **Figure 1** have been previously presented in the papers [8–15]. It should be remarked that the polarizability of the sensitized organic composites is larger than the one for the pure polymer or the liquid crystal matrix due to the formation of the larger dipole moment under the sensitization process.

The basic features of carbon nanotubes are regarded to their high conductivity, strong hardness of their C—C bonds, complicated and unique mechanisms of charge carrier moving, and large surface energy [16–22]. Moreover, the CNTs refractive index *n* is so small, namely, it is placed in the range of 1.05–1.1. Different types of the optical materials, especially inorganic ones, can be treated by CDV, PDV, laser ablation, laser oriented deposition, etc. technical methods, to deposit the CNTs on the materials surfaces in order to develop the novel coatings. Indeed, it can provoke the change of the refractivity, spectral, mechanical, and wetting phenomena of the composites. Analytical and quantum chemical simulations have supported the experimental results. Some accent has been given to form the covalent bonding between the atoms of carbon nanotubes and near-surface atoms of the matrix materials, see **Figure 2**. It provokes the dramatic increase of the transparency, mechanical hardness, laser strength, wetting angle, etc. parameters [23–27].

In the current "Fullerenes and relative materials" book, namely eight chapters are placed. They are devoted to study the fullerenes C_{60} and C_{70}, as well as the polymeric phase of fullerenes and their derivatives under high pressure and high temperature. The interesting results are presented about the possibility to encapsulate of 3–10 nitrogen atoms into the C_{70} cage.

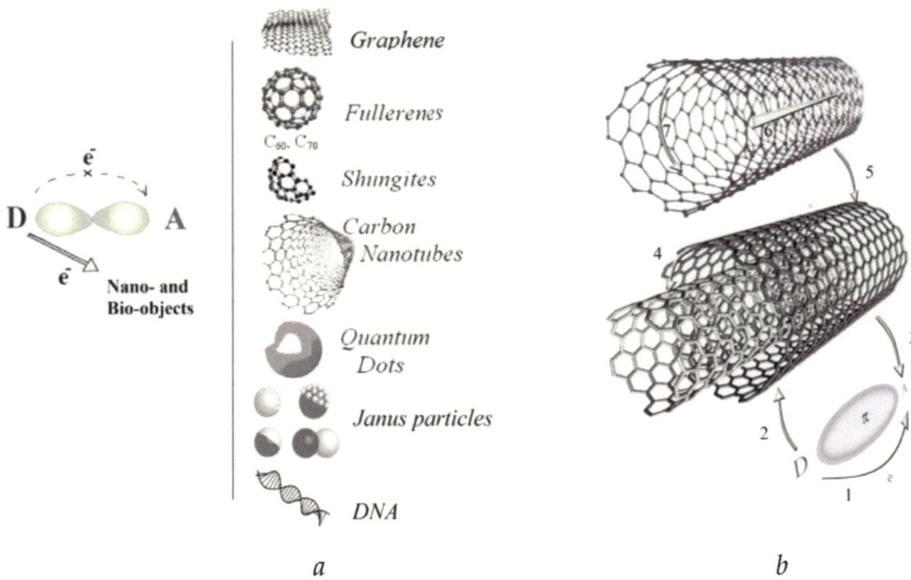

Figure 1. The possible mechanisms of CTC formation in the nano- and bio-objects-doped conjugated organics (a) with the selection of possible charge transfer on the CNTs basis (b).

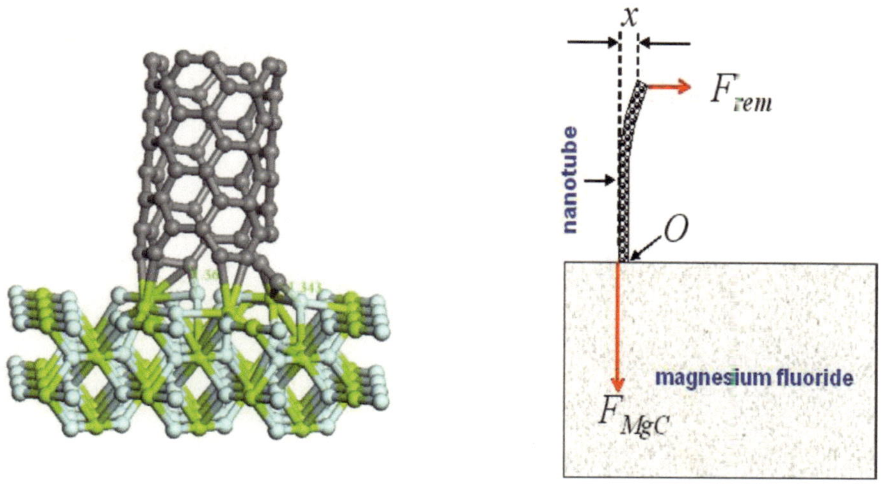

a b

Figure 2. Possible covalent bonding formation in the systems of MgF_2-CNTs (a) [24] and analytical calculation of this process (b) [25].

Raman spectrum of hydrogenated carbon film deposited by dc-pulse plasma CVD has been analyzed, and the process of the hybridization of graphene fragment, which limits the nucleation and growth of onion-like nanoparticle, has been shown. Optical, nonlinear optical, and mechanical features of the nano-objects containing composites are presented and supported by the analytical and quantum-chemical simulations.

All chapters present the results about the unique properties of the materials studied in the different area of their applications, including the general optoelectronics, solar energy and gas storage, laser and display, biomedicine, etc. It is important for the education process and for the civil and special device operation. The advanced idea, the special approach, and the information described in the current book will be fruitful for the readers to find a sustainable solution in a fundamental study and in the technique. The book can be useful for the students, post-graduate students, engineers, researchers, and technical officers of optoelectronic universities and companies.

Acknowledgements

The editor would like to thank all authors of the chapters presented, reviewers and all who have helped to prepare this book. The editor would like to acknowledge to Mr. Slobodan Momcilovic, Publishing Process Manager, InTech Publisher, Croatia for his good and continued cooperation.

Conflict of interest

There is no conflict of interest.

Author details

Natalia Vladimirovna Kamanina[1,2*]

*Address all correspondence to: nvkamanina@mail.ru

1 Vavilov State Optical Institute, Saint-Petersburg, Russia

2 St. Petersburg Electrotechnical University ("LETI"), Saint-Petersburg, Russia

References

[1] Kroto HW, Heath JR, O'Brien SC, Curl RF, Smalley RE. C_{60}: Buckminsterfullerene. Nature. 1985;**318**:162-163. DOI: 10.1038/318162a0

[2] Krätschmer W, Fostiropoulos K, Huffman DR. The infrared and ultraviolet absorption spectra of laboratory-produced carbon dust: Evidence for the presence of the C_{60} molecule. Chemical Physics Letters. 1990;**170**(2-3):167-170

[3] Krätschmer W, Lamb LD, Fostiropoulos K, Huffman DR. Solid C_{60}: A new form of carbon. Nature. 1990;**347**:354-358

[4] McKenzie DR, Davis CA, Cockayne DJH, Muller DA, Vassallo AM. The structure of the C_{60} molecules. Nature. 1992;**355**:622-624

[5] Couris S, Koudoumas E, Ruth AA, Leach S. Concentration and wavelength dependence of the effective third-order susceptibility and optical limiting of C_{60} in toluene solution. Journal of Physics B: Atomic, Molecular and Optical Physics. 1995;**8**:4537-4554

[6] Krieng T, Petr A, Barkleit G, Dunsch L. Improved thermal stability of nonpolymeric organic glasses by doping with fullerene C60. Applied Physics Letters. 1999;**74**(24):3639-3641

[7] Shibata T, Ishii T, Nozawa H, Tamamura T. High-aspect-ratio nanometer-pattern fabrication using fullerene-incorporated nanocomposite resists for dry-etching application. Japanese Journal of Applied Physics. 1997;**36**(12B):7642-7645

[8] Kamanina NV. Peculiarities of optical limiting effect in π-conjugated organic systems based on 2-cyclooctylamino-5-nitropyridinedoped with C_{70}. Journal of Optics A: Pure and Applied Optics. 2001;**3**(5):321-325. DOI: 10.1088/1464-4258/3/5/302

[9] Kamanina NV. Optical investigations of a C_{70}-doped 2-cyclooctylamino-5-nitropyridine-liquid crystal system. Journal of Optics A: Pure and Applied Optics. 2002;**4**(4):571-574. DOI: 10.1088/1464-4258/4/5/313

[10] Kamanina NV. Fullerene-dispersed liquid crystal structure: Dynamic characteristics and self-organization processes. Physics-Uspekhi. 2005;**48**(4):419-427. DOI: 10.1070/PU2005v048n04ABEH002101

[11] Kamanina NV, Emandi A, Kajzar F, Attias A-J. Laser-induced change in the refractive index in the systems based on nanostructured polyimide: Comparative study with other photosensitive structures. Molecular Crystals and Liquid Crystals. 2008;**486**:1-11

[12] Kamanina NV, Uskokovic DP. Refractive index of organic systems doped with nano-objects. Materials and Manufacturing Processes. 2008;**23**:552-556

[13] Kamanina NV, Serov SV, Shurpo NA, Likhomanova SV, Timonin DN, Kuzhakov PV, Rozhkova NN, Kityk IV, Plucinski KJ, Uskokovic DP. Polyimide-fullerene nanostructured materials for nonlinear optics and solar energy applications. Journal of Materials Science: Materials in Electronics. 2012;**23**(8):1538-1542. DOI: 10.1007/s10854-012-0625-9

[14] Kamanina NV, Zubtcova YuA, Kukharchik AA, Lazar C, Rau I. Control of the IR-spectral shift via modification of the surface relief between the liquid crystal matrixes doped with the lanthanide nanoparticles and the solid substrate. Optics Express. 2016;**24**(2):6 pages. DOI: 10.1364/OE.24.00A270. OPTICS EXPRESS A270

[15] Kamanina Natalia V, Serov Sergey V, Bretonniere Y, Ch A. Organic systems and their photorefractive properties under the nano- and biostructuration: Scientific view and sustainable development. Journal of Nanomaterials. 2015: Article ID 278902, 5 pages. DOI: 10.1155/2015/278902

[16] Namilae S, Chandra N, Shet C. Mechanical behavior of functionalized nanotubes. Chemical Physics Letters. 2004;**387**:247-252

[17] Fa W, Yang X, Chen J, Dong J. Optical properties of the semiconductor carbon nanotube intramolecular junctions. Physics Letters. 2004;**A323**:122-131 www.elsevier.com/locate/pla

[18] Taherpour AA, Aghagolnezhad-Gerdroudbari A, Rafiei S. Theoretical and quantitative structural relationship studies of reorganization energies of [SWCNT(5,5)-Armchair-CnH20] (n = 20-310) nanostructures by neural network CFFBP method. International Journal of Electrochemical Science. 2012;**7**:2468-2486 www.electrochemsci.org

[19] Yang Z-P, Ci L, Bur JA, Lin S-Y, Ajayan PM. Experimental observation of an extremely dark material made by a low-density nanotube array. Nano Letters. 2008;**8**(2):446-451. DOI: 10.1021/nl072369t

[20] Bao H, Ruan X, Fisher TS. Optical properties of ordered vertical arrays of multi-walled carbon nanotubes from FDTD simulations. Optics Express. 2010;**18**(6):6347-6359

[21] Chen K-R, Yeh H-F, Chen H-C, Liu T-J, Huang S-J, Ping-Yao W, Tiu C. Optical-electronic properties of carbon-nanotubes based transparent conducting films. Advances in Chemical Engineering and Science. 2013;**3**:105-111. DOI: 10.4236/aces.2013.31013

[22] Kvashnin DG, Sorokin PB. Effect of ultrahigh stiffness of defective graphene from atomistic point of view. Journal of Physical Chemistry Letters. 2015;**6**(12):2384-2387

[23] Kamanina NV, PYa V, Studeonov VI, Usanov YE. Strengthening transparent conductive coatings and "soft" materials of the IR range when nanotechnologies are used. Journal of Optical Technology. 2008;**75**(1):67-68

[24] Kamanina NV, Vasilyev PY, Studeonov VI. Using nanotechnologies in optics: On the possibility of enhancing the transparency and increasing the mechanical surface strength of materials in the UV and IR regions. Journal of Optical Technology. 2008;**75**(12):806-808

[25] Kamanina NV, Bogdanov KY, Vasilyev PY, Studeonov VI. Enhancing the mechanical surface strength of "soft" materials for the UV and IR ranges and increasing their transmission spectrum: Model MgF_2-nanotube system. Journal of Optical Technology. 2010;**77**(2):145-147

[26] Kamanina NV, Features of Optical Materials Modified with Effective Nanoobjects: Bulk Properties and Interface, New York, Physics Research and Technology, "Novinka". New York: Nova Science Publishers, Inc.; 2014. 116 p. ISBN: 978-1-62948-033-6

[27] Kamanina NV. Chapter 8. Perspective of the structuration process use in the optoelectronics, solar energy, and biomedicine. In: Vakhrushev A, editor. Nanomechanics. Rijeka: InTech; 2017. pp. 167-183, 192. ISBN 978-953-51-3182-3. DOI: 10.5772/68123

Section 1

On Synthesis Aspects

Chapter 2

Rational Synthesis of Fullerenes

Konstantin Amsharov

Additional information is available at the end of the chapter

http://dx.doi.org/10.5772/intechopen.73251

Abstract

Fullerenes are a unique family of carbon-based cage molecules, which attract interest because of their remarkable properties and potential applications. Most effort so far has been focused on the study of C_{60} and C_{70}, whereas other members of the huge fullerene family remain poorly explored. One of the main challenges in this field is the developing of the synthetic methods, which are suitable for the production of these unique materials in isomer-pure form in macroscopic amounts. Here, we review studies toward the rational synthesis of fullerenes from molecular precursors that have been published to date. The scope and limitation of the zipping strategy are discussed. The relevance and prospects for construction of the fullerene cages and related carbon-based nanostructures via cyclodehydrofluorination (C—F bond activation) are highlighted.

Keywords: fullerenes, bottom-up synthesis, zipping approach, C—F bond activation

1. Introduction

Fullerenes represent a unique class of sp^2-carbon-based compounds with a spherical π-system, displaying intriguing physical and chemical properties [1, 2]. The conventional methods of fullerene production are based on various graphite evaporation techniques which result in the heterogeneous fullerene containing soot [1, 2]. Unfortunately, none of the existing techniques provide access to fullerenes in isomer-pure form, which is crucial for their possible applications. The random nature of the fullerene formation results in a highly heterogeneous mixture containing mainly the most stable C_{60}-I_h and C_{70}-D_{5h} cages. Both fullerenes can be rather easily obtained in pure form by single-step HPLC separation of the fullerene soot extract (**Figure 1**). So far, only these two fullerenes have been the subject of in-depth investigations although the fullerene family represents a huge class of carbon-cage molecules. As soon as C_{60} became available in bulk amounts, a number of unique properties such as superconductivity [3], ferromagnetism [4, 5],

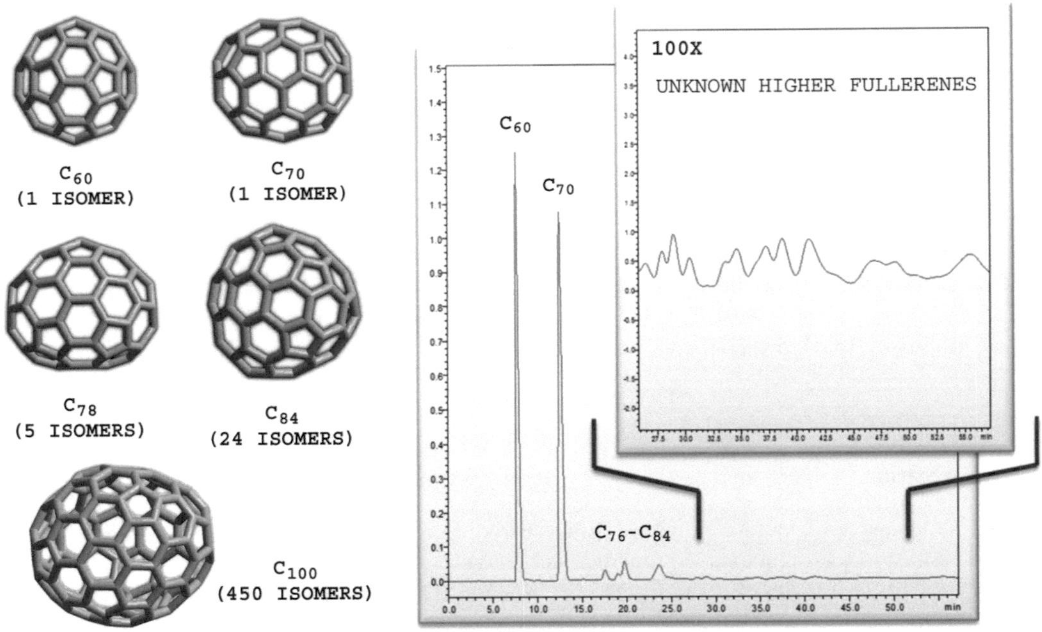

Figure 1. (Left) Several members of the fullerene family and the number of possible IPR isomers. (Right) Typical HPLC profile of the fullerene soot extract (Buckyprep column, toluene as eluent, UV detection). Two strong signals at 8 and 10 min correspond to C_{60}-I_h and C_{70}-D_{5h} fullerenes, which can be easily isolated. The fraction eluting between 17 and 25 min corresponds to abundant higher fullerenes—C_{76}-C_{84} (mixture of about 30 fullerene species). The huge number of small overlapping peaks (t_R > 28 min) corresponds to the mixture of a large number of unknown higher fullerenes.

and high electron affinity [6] have been disclosed. C_{60} and C_{70} fullerenes were found to be ideal material for the development of solar-energy conversion systems, in which the fullerene plays the role of effective electron acceptor [7]. Besides these two fullerene members, the fullerene soot typically contains about 1% of the so-called higher fullerenes (C2n with n > 35). In contrast to C_{60} and C_{70}, the higher fullerenes have several structural isomers obeying the isolated pentagon rule (IPR), stating that all pentagons have to be completely surrounded by hexagons [8]. Some of isomers of C_{76}-C_{84} can be obtained in small amounts by multi-step HPLC using recycling HPLC. However, the HPLC technique reaches its limit already on the separation of individual isomers of C_{84}. The separation of isomers of higher fullerenes with more than 84 carbon atoms is extremely complicated and these fullerenes remain highly unexplored (**Figure 1**). Moreover, the number of possible IPR isomers surges exponentially with increasing number of carbon atoms in the cage, which make the separation of higher fullerenes practically impossible. For example, C_{100} has 450 IPR isomers whereas for C_{200} the number of possible IPR isomers is 15.655.672. Furthermore, fullerenes that satisfy the IPR rule represent only a tiny fraction of possible carbon cages that in principle can be formed. Thus, C_{60} has 1811 non-IPR isomers whereas C_{200} has already more than 214 millions of isomers [8]. The non-IPR cages are remarkably less stable than IPR fullerenes, since adjacency of two pentagons resulted in high local strain and therefore high reactivity of these species. Nevertheless, it has been demonstrated that non-IPR cages exist and can be isolated at least in the form of exohedral or endohedral derivatives [9]. Furthermore, the possibility to introduce heteroatoms (heterofullerenes) [10], and/or to incorporate rings of different sizes (non-classical fullerenes) into the carbon cage [11], provides virtually unlimited number of possible fullerene structures. Each individual fullerene molecule possesses its own set of unique properties such as HOMO-LUMO gap, electron affinity, optical and magnetic characteristics. Despite intense research in the fullerene field during last decades, only a small fraction of possible fullerenes has been isolated in pure form and characterized. The higher fullerenes, as well as non-IPR and non-classical cages remain poorly explored due to the extremely low content in the soot and difficulties connected with purification. Further development in this field requires alternative synthetic methods which are suitable for the macroscopic production of desired fullerenes in isomer-pure form.

2. The concept of the rational synthesis of fullerenes

Rational synthesis of fullerenes is of practical interest as a promising method for the synthesis of isomer-pure fullerenes, including fullerenes which cannot be obtained by conventional techniques. One of the first attempts of the "rational synthesis" of fullerenes was undertaken in 1993 [12]. It has been proposed that C_{60}-I_h fullerene cage can be constructed by the fusion of six naphthalene moieties. Indeed, small amounts of C_{60} were detected in the pyrolysis products of the naphthalene performed in the argon atmosphere at 1000°C [12]. The proposed mechanism suggests a step-by-step assembly of naphthalene molecules via cyclodehydrogenation leading to various oligo-naphthalenes. Some of hexamers with defined orientation of naphthalene fragments possess the required for C_{60} connectivity, and thus can act as a precursor of C_{60}. The intramolecular cyclization of such precursors should unambiguously lead to the C_{60} cage formation. One of the possible naphthalene hexamers ($C_{60}H_{28}$), which is likely to form C_{60} trough intramolecular condensation, is shown in **Figure 2**. The low yield of C_{60} in this case might be explained by the low probability of the formation of the naphthalene hexamer with required orientation of subunits during accidental condensation. Indeed, as it was shown later, the pyrolysis of properly connected oligo-naphthalene improves the C_{60} yield [13]. In spite of the fact that this hypothesis does not explain the simultaneous formation of C_{70}, which was also observed in the pyrolysis products (direct assembly of C_{70} from naphthalene fragments is impossible), the given mechanism suggests that fullerenes can be built from well-defined aromatic fragments—the so-called fullerene precursors. These polycyclic aromatic hydrocarbons (PAHs), already containing required carbon connectivity, appear to be the key intermediates which exhaustive cyclization should lead to the strictly predefined carbon cage. In 2001, this methodology was successfully employed by Scott et al. for C_{60} fullerene synthesis under laser ablation conditions [14]. The rigid C_{60} fullerene precursor, containing all 60 carbon atoms at required positions, was obtained through a multi-step organic synthesis and subjected to the LDI-induced "pyrolysis" (laser desorption ionization). According to MS analysis the LDI-induced cyclization led the formation of C_{60} (**Figure 2**) [14]. Although the cyclization was found to be rather inefficient, the demonstrated principal transformation of the preprogrammed PAH precursor to the desired carbon cage paves the way for the rational construction of the fullerenes and other related carbon-based nanostructures.

The general strategy of the rational synthesis of fullerenes is based on the synthesis of specially "designed" PAH-precursor molecules containing the carbon framework required for

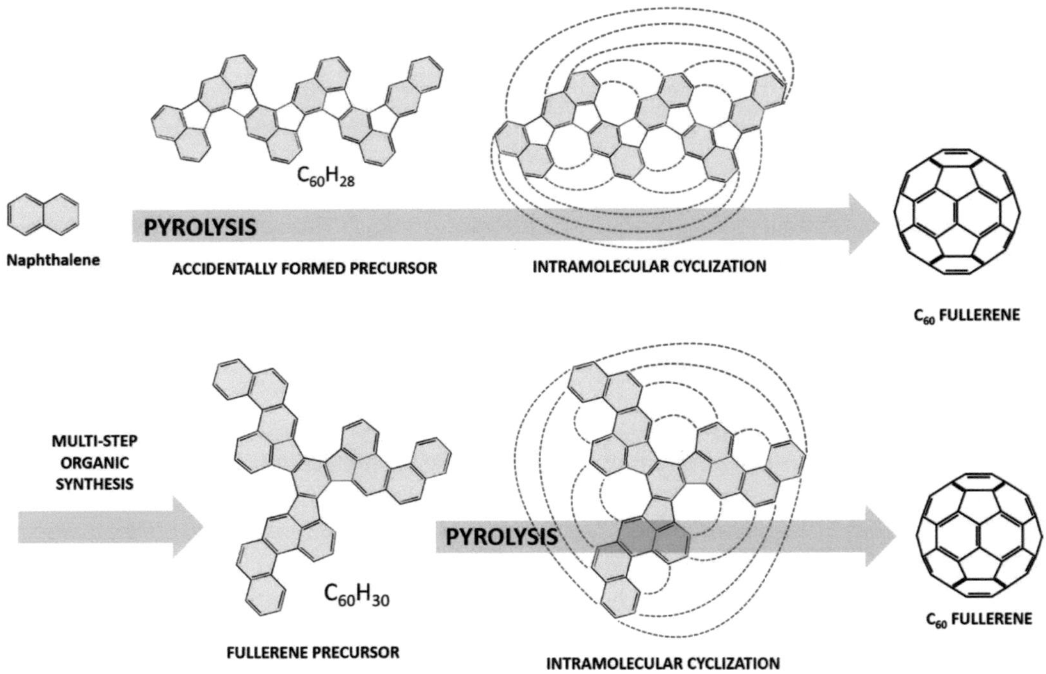

Figure 2. (Top) Schematic presentation of the C_{60} fullerene formation during naphthalene pyrolysis showing formation of the naphthalene hexamer as a key intermediate. (Bottom) Synthesis of C_{60} fullerene by zipping of the rigid atomically precise precursor under LDI conditions. The new C—C bond, which forms during cyclization, is coded by dashed lines.

the formation of the target fullerene cage. The respective precursor can be "zipped" to the desired fullerene by the tandem intramolecular Aryl—Aryl coupling. The characteristic feature of this approach is a zipper mechanism of cyclization in the final synthetic step. Since the regiospecificity of each condensation step can be strictly predefined by the precursor design, the cyclization will lead to the preprogrammed isomer of fullerene exclusively. This final cyclization step represents the key transformation in the rational synthesis of fullerenes by the zipping approach.

2.1. Wet-chemical synthesis

Since a big number of new C—C bonds have to be established during precursor zipping, the efficiency of each condensation has to be very high in order to achieve reasonable conversion. Unfortunately, methods developed for the synthesis of the planar PAHs via intramolecular Aryl–Aryl coupling are usually uneffective in the case of highly strained fullerenes. The strain energy of the fullerenes stems mainly due to pyramidalization of carbon atoms, as a result of deviation from energetically ideal values observed in graphene lattice. The perturbation of the π-system causes enormous enhancement of the energy, which has to be overcome during cyclization. Thus for example, various polyphenylenes can readily undergo an intramolecular cyclization under UV irradiation in the presence of an oxidant [15]. The photocyclization is an effective and frequently used technique for the synthesis of large planar polycyclic systems.

Alternatively, oxidative intramolecular cyclodehydrogenation can be carried out under Scholl condition [16]. Recently, this approach has become of high practical interest for the synthesis of large PAHs such as nanographenes and nanoribbons [17, 18]. Although both cyclodehydrogenation methods show excellent efficiency in the case of planar PAHs, these techniques appear to be not effective for the synthesis of bowl-shaped PAHs and fullerenes. Several alternative intermolecular aryl–aryl coupling techniques have been developed to overcome these disadvantages. Here the palladium(0)-catalyzed direct arylation utilizing bromo and chloro derivatives has been found to be effective for the synthesis of small fullerene fragments [19]. The largest fullerene fragments obtained so far contain 50 carbon atoms and formally represent more than 83% of the C_{60} fullerene surface (**Figure 3**) [20].

Although the largest synthesized C_{50} bowl is more curved than C_{60} [20], the rational synthesis of fullerene following this route still remains a challenge. This is mainly connected with difficulties to perform the Pd-catalyzed cyclization in a domino fashion. Secondly, the reaction conditions appear to be aggressive with respect to fullerene species. The recent discovery of effective intramolecular aryl–aryl coupling via C-F bond activation under transition-metal free conditions might be a solution to this problem. Recently, Siegel et al [21], Ichikawa et al. [22], and Amsharov et al. [23, 24] have demonstrated the efficiency of the C—F bond activation strategy for the synthesis of various bowl-shaped PAHs via formal HF elimination step (cyclodehydrofluorination). In the last case, the effective cyclization was achieved utilizing easily accessible aluminum oxide, which makes the approach highly attractive for preparative-scale synthesis. Moreover, it has been found that intramolecular Aryl-Aryl coupling of fluoroarenes via γ-aluminum oxide-mediated C—F bond activation can be realized under mild conditions with unprecedentedly high chemoselectivity [25, 26]. Using this technique an extended bowl-shaped PAHs representing more than 75% of the C_{60} fullerene connectivity was obtained with close to quantitative yield (**Figure 4**) [24], demonstrating high potential for generating extended nonplanar carbon-based nanostructures including fullerenes. Although the rather big fullerene fragments can be successfully produced by wet-chemical methods, the solution-synthesis of fullerenes remains a challenge.

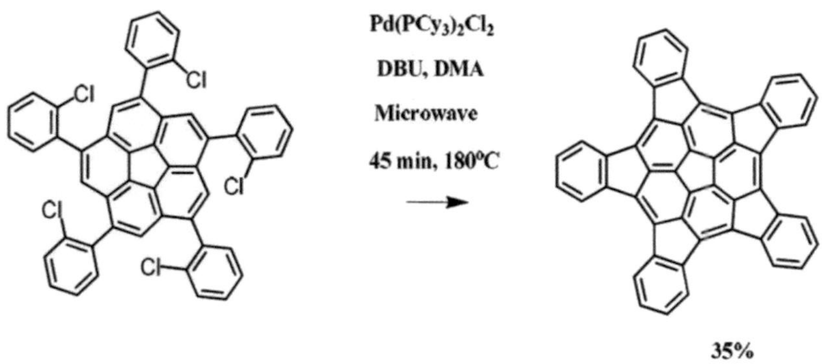

Figure 3. The synthesis of bowl-shaped PAH (C_{60} fullerene fragment) by Pd-catalyzed microwave-assisted intramolecular arylations.

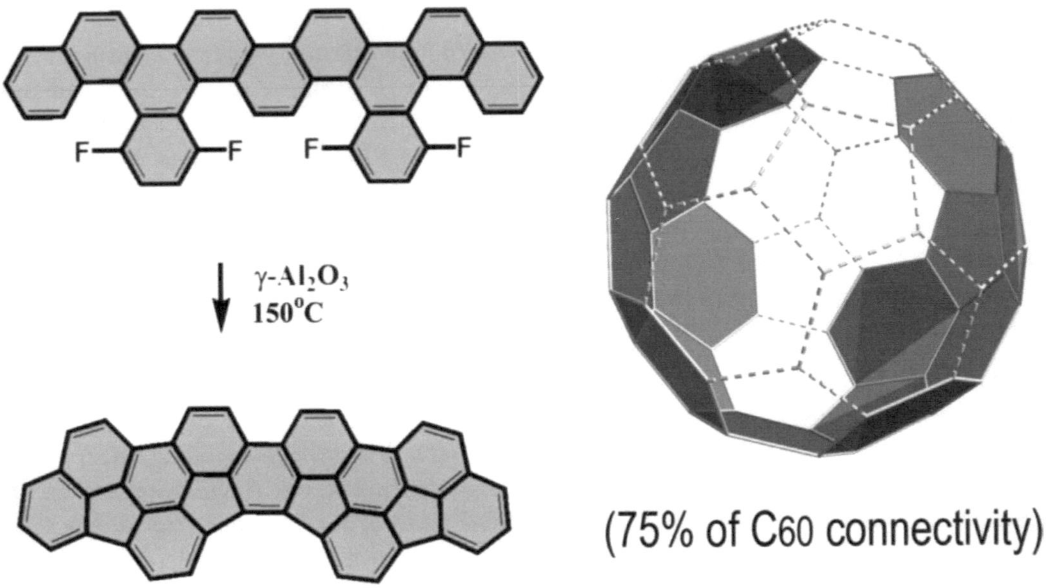

Figure 4. (Left) Synthesis of the extended fullerene fragments via alumina-assisted cyclodehydrofluorination. (Right) Superposition of C_{60} fullerene (dashed lines) and buckybowl (highlighted by gray) structures showing the same atom topology.

2.2. Surface-assisted synthesis

The direct on-surface synthesis of nanostructures is a highly promising strategy. The on-surface studies are usually conducted on single-crystal surfaces under ultra-high vacuum conditions that offer extraordinary degree of control in the experiments [27]. Moreover, the usage of atomically clean and flat crystal surfaces allows the observation of the reaction process at the single molecule level by means of scanning tunneling microscopy (STM) [28]. Recent progress in this field demonstrates a high potential for the synthesis of complex architectures via Aryl-Aryl coupling of organic precursors on single-crystal metal surfaces [29, 30]. In the case of planar PAHs, the intramolecular cyclodehydrogenation can be performed effectively on Cu(111), Ag(111), and Au(111) surfaces. Intramolecular cyclization on Au(111) surfaces typically occurs at around 400°C [31]. On the more reactive copper surfaces, cyclization takes place earlier at 200°C [32]. However, attempts to perform a similar cyclization to form non-planar PAHs were unsuccessful because of the intermolecular merging of the precursor molecules. Therefore, the synthesis of fullerenes requires harder conditions, including using more reactive Pt-group metal surfaces and applying higher temperatures of 450–600°C [33]. The first surface-assisted synthesis of the bowl-shaped PAH was reported by Rim et al. on the example of the hemispherical PAH C_{48}, which was generated on the Ru(0001) surface via exhaustive cyclodehydrogenation of hexabenzocoronene precursor (**Figure 5**) [34]. The six-fold cyclodehydrogenation on the periphery of precursor was achieved by annealing of the precursor molecule at 600°C. Note that Ru surface serves not only as a catalyst for the cyclodehydrogenation it also prevents intermolecular coupling due to efficient bonding to the surface. Finally, the Ru surface acts as a support for precursors and products enabling STM investigation. The STM analysis has

http://dx.doi.org/10.5772/intechopen.73251

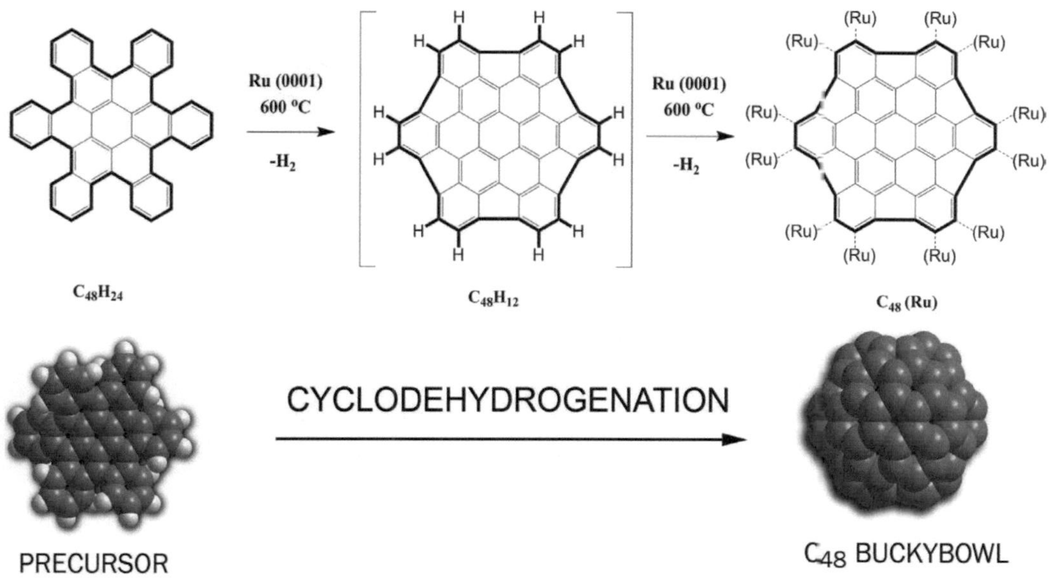

Figure 5. Surface-assisted synthesis of the hemispherical C_{48} buckybowl via exhaustive cyclodehydrogenation of hexabenzocoronene on Ru surface at 600°C.

revealed a high selectivity and exceptionally high conversion ratio of the deposited precursors into non-planar PAH structure (**Figure 5**). However, due to high reactivity of the Ru surface and high temperature applied, the dehydrogenation process does not stop on the formation of the $C_{48}H_{12}$ bowl. The bowl-shaped PAH undergoes further dehydrogenation resulting in all-carbon C_{48} bowl covalently bonded to the Ru surface by 12 C—Ru bonds (**Figure 5**) [34].

Soon after, the efficiency of the surface-assisted approach has been demonstrated by Otero et al. on the synthesis of C_{60} fullerene on Pt surface [35]. In their study, $C_{60}H_{30}$ fullerene precursor molecules (**Figure 6**, precursor A) were deposited onto a Pt(111) surface and annealed at around 480°C. The annealing step led to the formation of C_{60} cages with an efficiency close to 100%. The STM analysis shows that the molecular coverage does not change during temperature-induced reaction, and all of the precursor molecules converted into fullerenes [35]. Later it was shown that the process is highly selective in nature and is not accomplished by skeletal rearrangements. This important question has been addressed by the investigation of surface-catalyzed cyclodehydrogenation of a specifically designed precursor molecule (**Figure 6**, precursor B) "preprogrammed" to the formation of the C_{60}-bowl-shaped PAH with unique triangular shape [36]. The on-surface cyclodehydrogenation of two isomeric $C_{60}H_{30}$ precursors (**Figure 7**) gave convincing evidence that the reaction occurs in a highly selective manner [36]. In contrast to the condensation of the precursor A whose cyclodehydrogenation leads to the spherical C_{60}, the zipping of the precursor B resulted in a triangular-shaped C_{60}-bowls whose formation was distinguished by the STM analysis.

Those works demonstrated that the surface-assisted cyclodehydrogenation on Pt-group metal surfaces is highly selective and can be used for fabrication of various specifically designed carbon-based nanostructures including fullerenes.

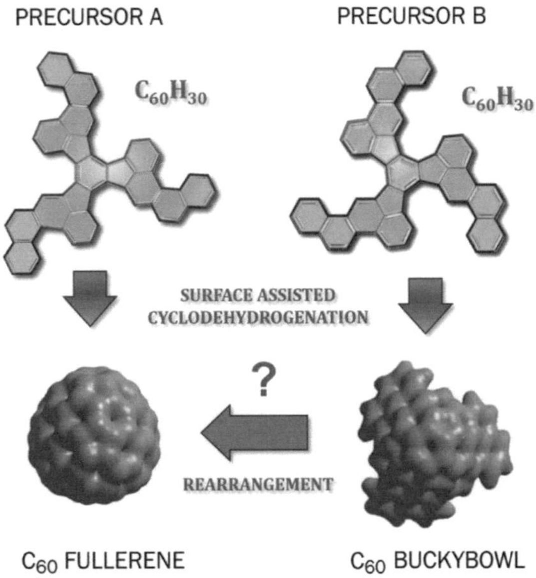

Figure 6. Structures of the C_{60} fullerene precursor (A and the C_{60} buckybowl precursor (B) molecules and their cyclodehydrogenation products.

Finally, the approach was successfully applied for the first rational synthesis of isomer-pure higher fullerene C_{84}(20) [36]. Upon annealing of specially designed $C_{84}H_{42}$ precursor at 550°C on Pt(111) surface, the shape of all of precursor molecules changed and all species became spherical with a lateral size expected for C_{84} fullerene (**Figure 8**). The formation of the C_{84} fullerene cages was additionally confirmed by MS analysis [36]. Moreover, the approach was found to be suitable for the synthesis of highly interesting hetero fullerenes. The fabrication of an azafullerene by surface-assisted approach has been demonstrated by Otero et al. on the example of $C_{57}N_3$ synthesis. In this study, a structurally related precursor with three incorporated nitrogen atoms was quantitatively converted to the hetero fullerene $C_{57}N_3$ upon annealing on Pt(111) surface at 477°C [35].

Although the technique allows to perform effective zipping to the target carbon nanostructures [37, 38], the resulting nano-carbons are covalently bonded to the metal surface and cannot be desorbed without decomposition. Those difficulties could be probably overcome by using less active metal oxide surfaces. Since the metal oxides show low activity in cyclodehydrogenation, the cyclization on such surfaces would require much higher temperatures leading to loss of selectivity. However, introduction of "activating" groups into the precursor structures could significantly reduce the activation barriers. In particular, it has been found that fluorine functionality appears to be an excellent ring-closure promoter, which has demonstrated unexpectedly high efficiency. It has been found that coupling can be realized in fluoroarenes on thermally activated aluminum oxide at very mild conditions [23–26]. Moreover, the alumina-supported cyclodehydrofluorination (cyclization via formal HF elimination) has demonstrated unprecedentedly high chemoselectivity and regiospecificity. The intramolecular Aryl–Aryl coupling on aluminum oxide was achieved with close to quantitative yield even

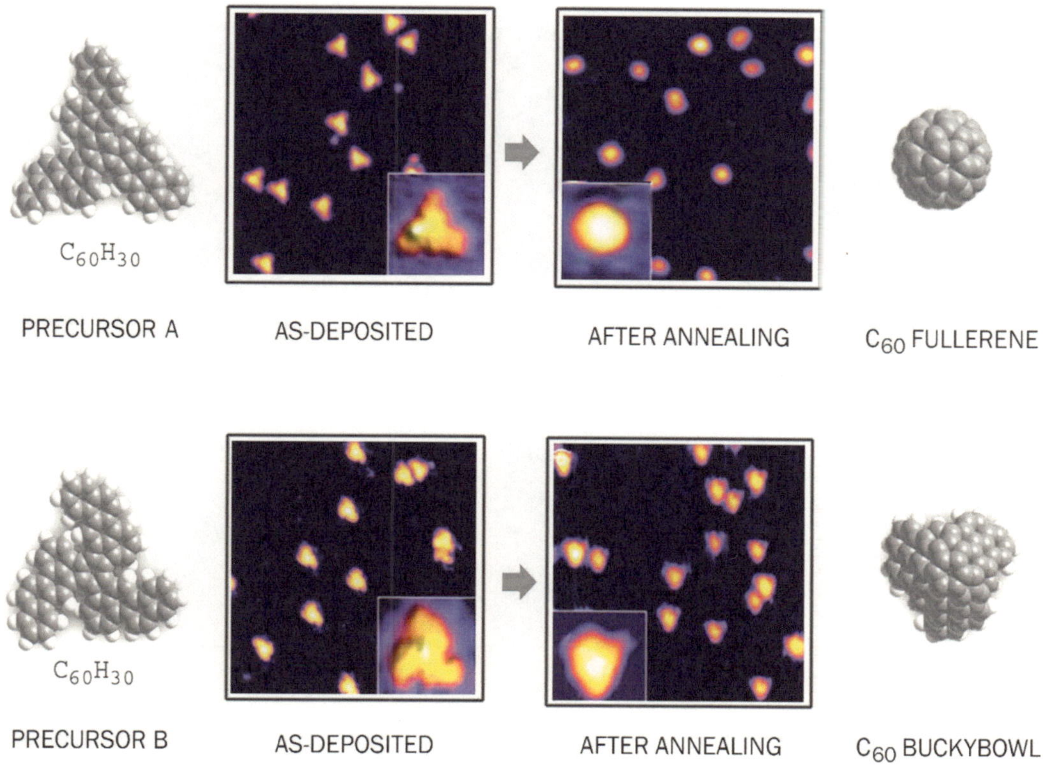

Figure 7. Surface-assisted synthesis of fullerene C_{60}-I_h (top) and C_{60} buckybowl (bottom) via cyclodehydrogenation of respective precursors on Pt(111) surface. STM images of the precursor before and after annealing at 480°C (reprint with permission from Ref [36]).

in the case of large bowl-shaped PAHs. Since fluorine can promote the ring closure only if hydrogen is placed neighboring in space in the precursor structure, it was possible to perform the cyclization in domino fashion. This makes the cyclodehydrofluorination more attractive in comparison to cyclodehydrogenation where the selectivity must be predefined by the rigid structure of precursor. Further, it was found that metal oxides of II, III and IV groups are active in the cyclodehydrofluorination of fluoroarenes. In particular, zirconium, titanium and indium oxides were found to be effective for cyclization via C—F bond activation [39]. These results point the way for fabrication of higher fullerenes and other related carbon-based nanostructures in a fully controllable manner directly on metal oxide surfaces.

2.3. Gas phase synthesis

The rational construction of fullerenes in the gas phase seems to be very logic, since in all conventional techniques the cage fabrication occurs in the gas phase. First of all, fullerenes, including higher, non-IPR and heterofullerenes, are stable in the gas phase. Second, the undesired intermolecular interactions can be effectively suppressed in the gas phase, providing excellent conditions for intramolecular processes. Not surprising that gas-phase synthesis appears to be the most promising approach.

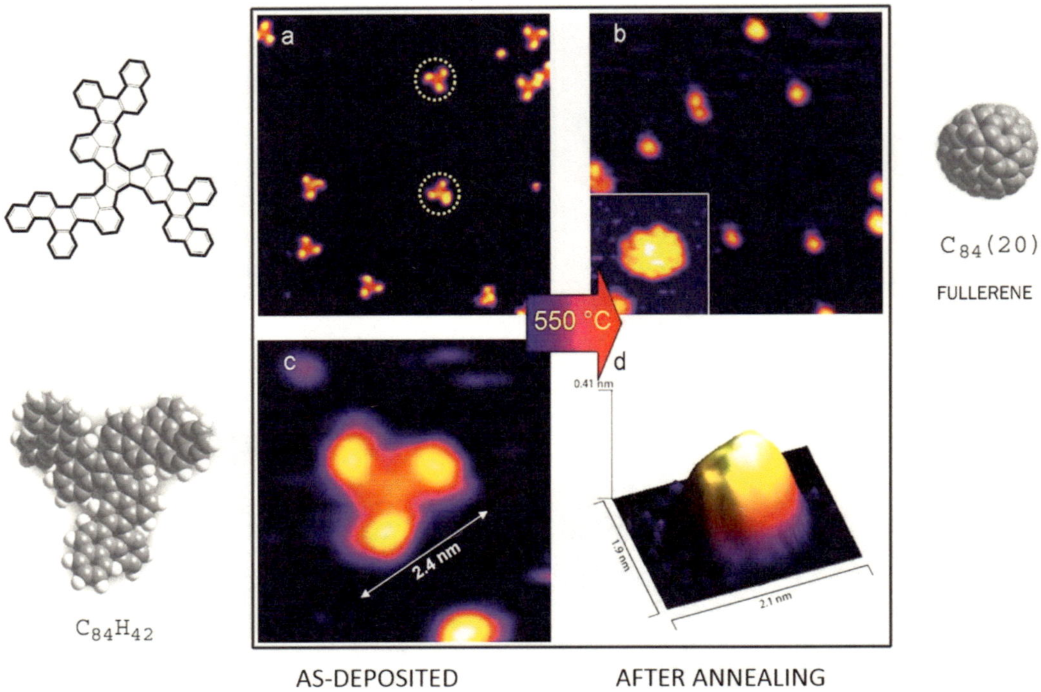

Figure 8. Surface-assisted synthesis of fullerene $C_{84}(20)$ via cyclodehydrogenation of $C_{84}H_{42}$ precursors on Pt(111) surface. STM images of the precursor before and after annealing at 550°C (reprint with permission from Ref. [36]).

The flash vacuum pyrolysis (FVP) represents one of the most powerful techniques for the synthesis of highly strained architectures via intramolecular cyclization. It consists in the heating of precursor molecules in the gas phase to a very high temperature for a very short time. The high temperature applied in FVP causes hemolytic cleavage of C—H bonds leading to the formation of radicals possessing enough energy to cause intramolecular cyclization yielding highly strained products. Pioneering by Scott et al., variety of small bowl-shaped PAHs (fullerene fragments) have been successfully synthesized in macroscopic amounts via intramolecular cyclization of quasi-planar PAH precursors under FVP conditions [40–42]. It was found that introduction of chlorine or bromine functionalities into the initial precursor is essential to facilitate the cyclization via free radical mechanism (**Figure 9**). Finally, this methodology has been successfully applied for the first rational synthesis of C_{60} utilizing partially chlorinated C_{60} fullerene precursor (**Figure 9**) [43]. Despite the low yield of C_{60} (0.1–1%) the power of the strategy has been clearly demonstrated.

The possibility of the synthesis of higher fullerenes by FVP has been shown on the examples of C_{78} [44] and C_{84} [45] by pyrolysis of specially programmed precursors. In both cases the size-selective formation of fullerenes was observed as indicated by the MS analysis (**Figure 10**). However, the rates of conversion to the target molecules have remained disappointingly low because of the low efficiency of intramolecular condensation of non-activated precursors. The usually employed for activation chlorine reach their limits in the case of large molecules because of decomposition of precursors during sublimation. The molar masses will generally

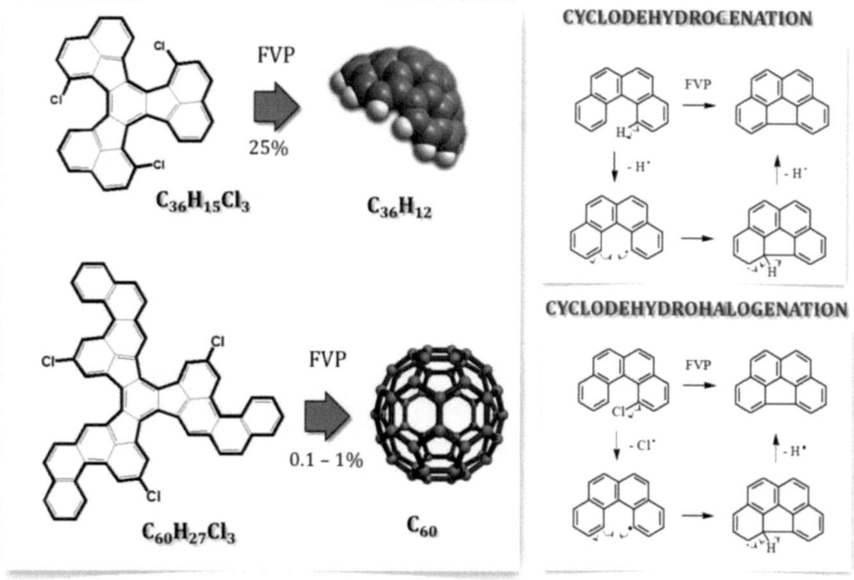

Figure 9. (Left) Synthesis of fullerenes and fullerene fragments by FVP approach from chlorinated PAH precursors. (Right) The cyclization mechanism under FVP condition for activated (chlorinated) and non-activated PAHs.

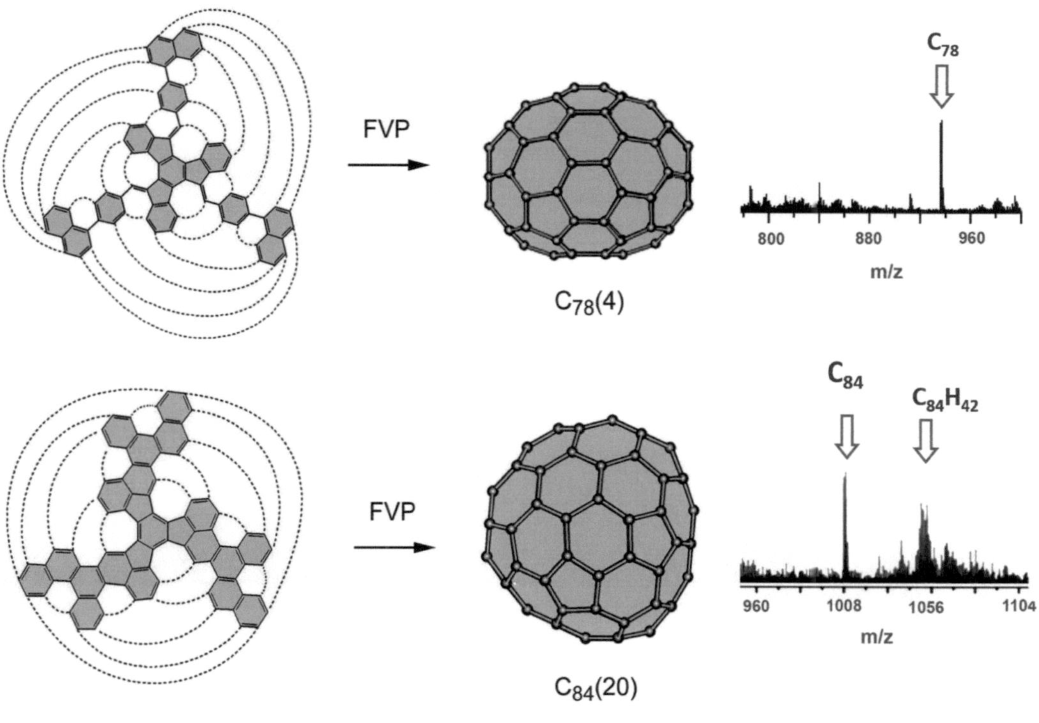

Figure 10. Schematic representation of the intramolecular cyclodehydrogenation of fullerene precursors to fullerenes $C_{78}(4)$ and $C_{84}(20)$. Dashed lines indicate where the new bonds have to be established in order to create a closed fullerene cage.

be high, since a large number of new C–C bonds that need to be formed necessitate the introduction of a big number of promoter groups. Moreover, the radical nature of the condensation could drastically affect the selectivity of the process.

Although the FVP approach has been proven to be very prolific for the synthesis of small non-planar PAHs, the high-yield synthesis of large bowls and isomerically pure fullerenes has remained challenging. Recently, an efficient intramolecular fluorine-promoted ring closure in benzo[c]phenanthrenes under FVP conditions via HF elimination has been reported [46]. It has been shown that HF elimination is a synchronous process leading directly to the C–C bond formation without any intermediates, thus producing no side products [46]. The small size and low molecular weight of fluorine, as well as high thermostability of the C–F bond make fluorine a perfect activating group for the rational synthesis of fullerene by zipping approach. The feasibility of the approach has been demonstrated on the successful synthesis of C_{60} fullerene from the respective fluorinated precursor which was effectively folded into the fullerene under laser ionization. The respective 15-fold cyclization was realized in a domino fashion via consecutive HF elimination without observation of side reactions or undesired fragmentations [47]. As it is shown in **Figure 11**, the precursor molecule undergoes highly selective HF loss (with simultaneous C–C bond formation) leading to the open-cage which

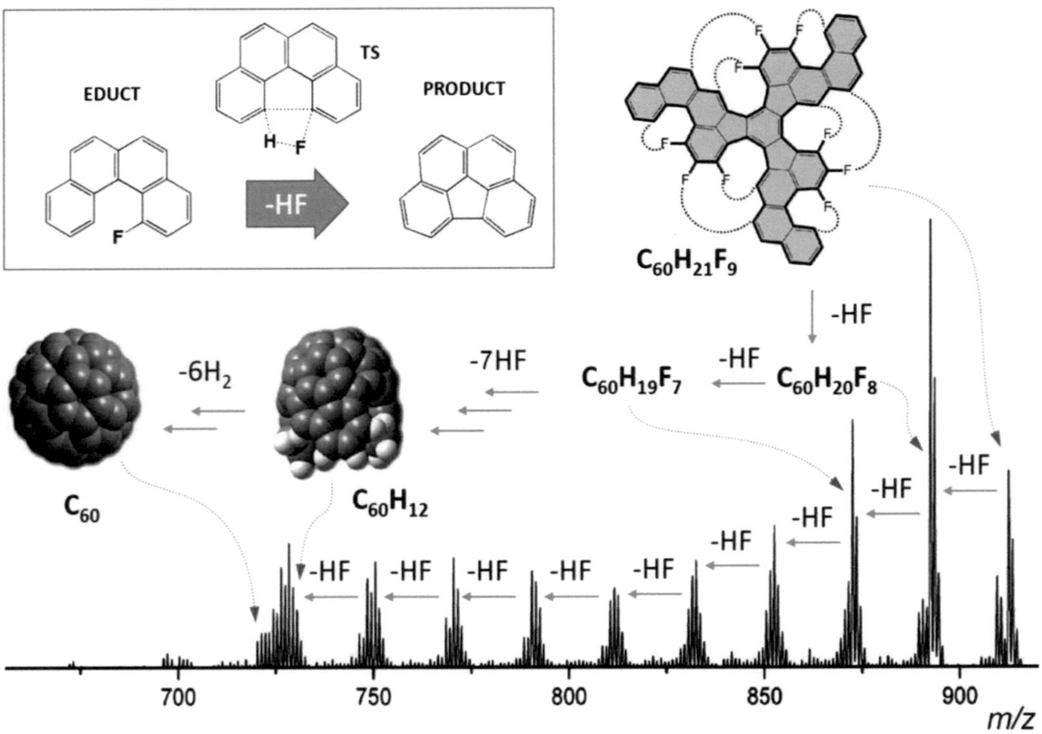

Figure 11. Mass spectrum of the C_{60} fullerene precursor ($C_{60}H_{21}F_9$) obtained under laser ablation (negative-ion mode) showing an effective domino-like HF elimination yielding C_{60} fullerene via 15 selective C–C bond formation steps. The mechanism of single cyclodehydrofluorination involving a four-centered transition state is shown in the insert on the example of benzo[c]phenanthrene model.

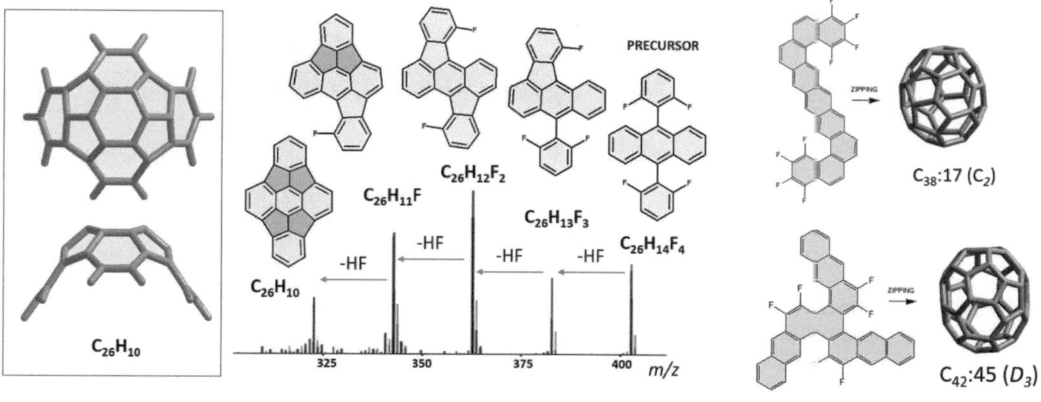

Figure 12. Mass spectrum of model precursor $C_{26}H_{14}F_4$ obtained under laser ablation showing tandem cyclodehydrofluorination yielding the non-IPR $C_{26}H_{10}$ fragment. The DFT optimized geometry of the non-IPR bowl (side and top views) is shown in the inset. (Right) Two examples of possible non-IPR fullerene precursor.

spontaneously collapses to the target C_{60} via H_2 loss. The regioselectivity of the domino-like cyclization and the formation of bowl-shaped intermediates and finally spherical C_{60}-I_h cage were recently confirmed by ion-mobility experiments [48].

Due to exceptionally high efficiency, it was assumed that the same approach can be extended to the synthesis of higher fullerenes as well as to the synthesis of highly attractive non-IPR cages. Recently, it has been demonstrated that non-IPR fragments can be indeed fabricated via cyclodehydrofluorination in the gas phase. It was shown that despite high strain the respective precursor undergoes rather smooth four-fold cyclodehydrofluorination, leading to the non-IPR fullerene fragment $C_{26}H_{10}$ (**Figure 12**) [49]. These results demonstrate the general possibility of rational fabrication of the non-IPR architectures via cyclodehydrofluorination of the specially preprogrammed fluoroarene-based precursors. Two possible precursors for generation of non-IPR fullerenes C_{38} and C_{42} are shown as example. Taking into account the present state-of-the-art of MS-based synthesis/deposition technique, which allows multilayer deposition of such compounds, the future scale-up appears to be feasible. Among non-IPR cages, the laser-induced cyclodehydrofluorination approach seems to be also applicable for the rational synthesis of higher and non-classical carbon cages.

3. Conclusion

The zipping strategy had been proved to be prolific for the rational synthesis of the fullerene fragments and the fullerene cages. The progress in this field shows the general possibility to fabricate elusive fullerenes in a fully controllable manner. Production of these unique materials in isomer-pure form in macroscopic amounts does no longer appear elusive, and there is promising prospect that such compounds will become available in the near future. Since there are virtually no limiting factors, there are no principle obstacles standing against extending the zipping approach to highly interesting related carbon-based architectures.

Acknowledgements

The author thanks the Deutsche Forschungsgemeinschaft (DFG-SFB 953 "Synthetic Carbon Allotropes").

Author details

Konstantin Amsharov

Address all correspondence to: konstantin.amsharov@fau.de

Department of Organic Chemistry, Friedrich Alexander University Erlangen-Nuremberg, Erlangen, Germany

References

[1] Hirsch A, Brettreich M. Fullerenes: Chemistry and Reactions. Weinheim, FRG: Wiley-VCH Verlag GmbH & Co. KGaA; 2004. 423 pp. DOI: 10.1002/3527603492

[2] Langa F, Nierengarten JF. Fullerenes: Principles and Applications. Cambridge: RSC Publishing; 2011. 634 pp. DOI: 10.1039/9781849732956

[3] Hebard AF, Rosseinsky MJ, Haddon RC, Murphy DW, Glarum SH, Palstra TTM, Ramirez AP, Kortan AR. Superconductivity at 18 K in potassium-doped C_{60}. Nature. 1991;**350**:600-601. DOI: 10.1038/350600a0

[4] Allemand PM, Khemani KC, Koch A, Wudl F, Holczer K, Donovan S, Gruner G, Thompson JD. Organic molecular soft ferromagnetism in a fullerene C_{60}. Science. 1991;**253**:301-302. DOI: 10.1126/science.253.5017.301

[5] Stephens PW, Cox D, Lauther JW, Mihaly L, Wiley JB, Allemand PM, Hirsch A, Holczer K, Li Q, Thompson JD, Wudl F. Lattice structure of the fullerene ferromagnet TDAE–C_{60}. Nature. 1992;**355**:331-332. DOI: 10.1038/355331a0

[6] Reed CA, Bolskar RD. Discrete fulleride anions and fullerenium cations. Chemical Reviews. 2000;**100**:1075-1119. DOI: 10.1021/cr980017o

[7] Lu L, Zheng T, Wu Q, Schneider AM, Zhao D, Yu L. Recent advances in bulk heterojunction polymer solar cells. Chemical Reviews. 2015;**115**:12666-12731. DOI: 10.1021/acs.chemrev.5b00098

[8] Fowler PW, Manolopoulos DE. An Atlas of Fullerenes. Oxford: Clarendon Press; 1995. 392 pp

[9] Tan YZ, Xie SY, Huang RB, Zheng LS. The stabilization of fused-pentagon fullerene molecules. Nature Chemistry. 2009;**1**:450-460. DOI: 10.1038/nchem.329

[10] Vostrowsky O, Hirsch A. Heterofullerenes. Chemical Reviews. 2006 **106**:5191-5207. DOI: 10.1021/cr050561e

[11] Gan LH, Zhao JQ, Hui Q. Nonclassical fullerenes with a heptagon violating the pentagon adjacency penalty rule. Journal of Computational Chemistry. 2010;**31**:1715-1721. DOI: 10.1002/jcc.21459

[12] Taylor R, Langley GJ, Kroto HW, Walton DRM. Formation of C_{60} by pyrolysis of naphthalene. Nature. 1993;**366**:728-731. DOI: 10.1038/366728a0

[13] Amsharov KY, Simeonov K, Jansen M. Formation of fullerenes by pyrolysis of 1,2′-binaphthyl and 1,3-oligonaphthylene. Carbon. 2007;**45**:337-343. DOI: 10.1016/j.carbon.2006.09.013

[14] Boorum MM, Vasil'ev YV, Drewello T, Scott LT. Groundwork for a rational synthesis of C_{60}: Cyclodehydrogenation of a $C_{60}H_{30}$ polyarene. Science. 2001;**294**:828-831. DOI: 10.1126/science.1064250

[15] Mallory FB, Mallory CW. Photocyclization of stilbenes and related molecules. Organic Reactions. Hoboken, New Jersey: John Wiley & Sons, Inc.; 2005. 456 pp. DOI: 10.1002/0471264180.or030.01

[16] Grzybowski M, Skonieczny K, Butenschon H, Gryko DT. Comparison of oxidative aromatic coupling and the Scholl reaction. Angewandte Chemie, International Edition. 2013;**52**:9900-9930. DOI: 10.1002/anie.201210238

[17] Chen L, Hernandez Y, Feng XL, Mullen K. From nanographene and graphene nanoribbons to graphene sheets: Chemical synthesis. Angewandte Chemie, International Edition. 2012;**51**:7640-7654. DOI: 10.1002/anie.201201084

[18] Kivala M, Wu D, Feng X, Li C, Mullen K. Cyclodehydrogenation in the synthesis of graphene-type molecules. In: Materials Science and Technology. Weinheim: Wiley-VCH; 2013. pp. 373-420. DOI: 10.1002/9783527603978.mst0423

[19] Alberico D, Scott ME, Lautens M. Aryl–aryl bond formation by transition-metal-catalyzed direct arylation. Chemical Reviews. 2007;**107**:174-238. DOI: 10.1021/cr0509760

[20] Steinberg BD, Jackson EA, Filatov AS, Wakamiya A, Petrukhina MA, Scott LT. Aromatic π-systems more curved than C_{60}. The complete family of all indenocorannulenes synthesized by iterative microwave-assisted Intramolecular arylations. Journal of the American Chemical Society. 2009;**131**:10537-10545. DOI: 10.1021/ja9031852

[21] Allemann O, Duttwyler S, Romanato P, Baldridge KK, Siegel JS. Proton-catalyzed, silane-fueled Friedel-Crafts coupling of fluoroarenes. Science. 2011;**332**:574-577. DOI: 10.1126/science.aaa4609

[22] Fuchibe K, Mayumi Y, Zhao N, Watanabe S, Yokota M, Ichikawa J. Domino synthesis of fluorine-substituted polycyclic aromatic hydrocarbons: 1,1-Difluoroallenes as synthetic platforms. Angewandte Chemie, International Edition. 2013;**52**:7825-7828. DOI: 10.1002/anie.201302740

[23] Amsharov KY, Merz P. Intramolecular aryl-aryl coupling of fluoroarenes through Al_2O_3-mediated HF elimination. The Journal of Organic Chemistry. 2012;**77**:5445-5448. DOI: 10.1021/jo300783y

[24] Amsharov KY, Kabdulov MA, Jansen M. Facile bucky-bowl synthesis by regiospecific Cove-region closure by HF elimination. Angewandte Chemie, International Edition. 2012;**51**:4594-4597. DOI: 10.1002/anie.201200516

[25] Papaianina O, Akhmetov VA, Goryunkov AA, Hampel F, Heinemann FW, Amsharov KY. Synthesis of rationally halogenated buckybowls via chemoselective aromatic C—F bond activation. Angewandte Chemie, International Edition. 2017;**56**:4834-4838. DOI: 10.1002/anie.201700814

[26] Steiner AK, Amsharov KY. The rolling-up of oligophenylenes to nanographenes by a HF-zipping approach. Angewandte Chemie, International Edition. 2017;**56**:14732-14736. DOI: 10.1002/anie.201707272

[27] Tait SL. Function follows form: Exploring two-dimensional supramolecular assembly at surfaces. ACS Nano. 2008;**2**:617-621. DOI: 10.1021/nn800207w

[28] Hla SW, Rieder KH. STM control of chemical reaction: Single-molecule synthesis. Annual Review of Physical Chemistry. 2003;**54**:307-330. DOI: 10.1146/annurev.physchem.54.011002.103852

[29] Gourdon A. On-surface covalent coupling in ultrahigh vacuum. Angewandte Chemie, International Edition. 2008;**47**:6950-6953. DOI: 10.1002/anie.200802229

[30] Mendez J, Lopez MF, Martin-Gago JA. On-surface synthesis of cyclic organic molecules. Chemical Society Reviews. 2011;**40**:4578-4590. DOI: 10.1039/C0CS00161A

[31] Cai J, Ruffieux P, Jaafar R, Bieri M, Braun T, Blankenburg S, Muoth M, Seitsonen AP, Saleh M, Feng X, Müllen K, Fasel R. Atomically precise bottom-up fabrication of graphene nanoribbons. Nature. 2010;**466**:470-473. DOI: 10.1038/nature09211

[32] Treier M, Pignedoli CA, Laino T, Rieger R, Müllen K, Passerone D, Fasel R. Surface-assisted cyclodehydrogenation provides a synthetic route towards easily processable and chemically tailored nanographenes. Nature Chemistry. 2011;**3**:61-67. DOI: 10.1038/nchem.891

[33] Amsharov K. Rational synthesis of fullerenes, buckybowls, and single-walled carbon nanotubes by a surface-assisted approach. Advances in polymer science. In: Müllen K, Feng X, editors. From Polyphenylenes to Nanographenes and Graphene Nanoribbons. Cham: Springer; 2017. pp. 127-145. DOI: 10.1007/12_2017_7

[34] Rim KT, Siaj M, Xiao SX, Myers M, Carpentier VD, Liu L, Su CC, Steigerwald ML, Hybertsen MS, McBreen PH, Flynn GW, Nuckolls C. Forming aromatic hemispheres on transition-metal surfaces. Angewandte Chemie, International Edition. 2007;**46**:7891-7895. DOI: 10.1002/anie.200701117

[35] Otero G, Biddau G, Sanchez-Sanchez C, Caillard R, Lopez MF, Rogero C, Palomares FJ, Cabello N, Basanta MA, Ortega J, Mendez J, Echavarren AM, Perez R, Gomez-Lor B, Martin-Gago JA. Fullerenes from aromatic precursors by surface-catalysed cyclodehydrogenation. Nature. 2008;**454**:865-868. DOI: 10.1038/nature07193

[36] Amsharov KY, Abdurakhmanova N, Stepanow S, Rauschenbach S, Jansen M, Kern K. Towards the isomer-specific synthesis of higher fullerenes and buckybowls by the surface-catalyzed cyclodehydrogenation of aromatic precursors. Angewandte Chemie, International Edition. 2010;**49**:9392-9396. DOI: 10.1002/anie.201005000

[37] Sanchez-Valencia JR, Dienel T, Groning O, Shorubalko I, Mueller A, Jansen M, Amsharov K, Ruffieux P, Fasel R. Controlled synthesis of single-chirality carbon nanotubes. Nature. 2014;**512**:61-64. DOI: 10.1038/nature13607

[38] Abdurakhmanova N, Mueller A, Stepanow S, Rauschenbach S, Jansen M, Kern K, Amsharov KY. Bottom up fabrication of (9,0) zigzag and (6,6) armchair carbon nanotube end-caps on the Rh(111) surface. Carbon. 2015;**84**:444-447. DOI: 10.1016/j.carbon.2014.12.038

[39] Amsharov KY. Cyclodehydrofluorination of fluoroarenes on metal oxides. Towards bottom-up synthesis of carbon nanostructures on insulating surfaces. Physica Status Solidi B: Basic Solid State Physics. 2016;**253**:2473-2477. DOI: 10.1002/pssb.201600211

[40] Scott LT. Methods for the chemical synthesis of fullerene. Angewandte Chemie, International Edition. 2004;**43**:4994-5007. DOI: 10.1002/anie.200400661

[41] Tsefrikas VM, Scott LT. Geodesic polyarenes by flash vacuum pyrolysis. Chemical Reviews. 2006;**106**:4868-4884. DOI: 10.1021/cr050553y

[42] Petrukhina MA, Scott LT. Fragments of Fullerenes and Carbon Nanotubes. Hoboken, NJ, USA: John Wiley & Sons, Inc.; 2012. 440 pp. DOI: 10.1002/9781118011263

[43] Scott LT, Boorum MM, McMahon BJ, Hagen S, Mack J, Blank J, Wegner H, de Meijere A. A rational chemical synthesis of C_{60}. Science. 2002;**295**:1500-1502. DOI: 10.1126/science.1068427

[44] Amsharov KY, Jansen M. A C_{78} fullerene precursor: Toward the direct synthesis of higher fullerenes. The Journal of Organic Chemistry. 2008;**73**(7):2931-2934. DOI: 10.1021/jo7027008

[45] Amsharov KY, Jansen M. Synthesis of a higher fullerene precursor—An "unrolled" C_{84} fullerene. Chemical Communications. 2009;(19):2691-2693. DOI: 10.1039/B901496A

[46] Amsharov KY, Kabdulov MA, Jansen M. Homo-elimination of HF—An efficient approach for intramolecular aryl-aryl coupling. Chemistry--A European Journal. 2010;**16**:5868-5871. DOI: 10.1002/chem.201000374

[47] Kabdulov M, Jansen M, Amsharov KY. Bottom-up C_{60} fullerene construction from a fluorinated $C_{60}H_{21}F_9$ precursor by laser-induced tandem cyclization. Chemistry--A European Journal. 2013;**19**:17262-17266. DOI: 10.1002/chem.201303838

[48] Greisch JF, Amsharov KY, Weippert J, Weis P, Böttcher A, Kappes MM. From planar to cage in 15 easy steps: Resolving the $C_{60}H_{21}F_9 \rightarrow C_{60}$-transformation by ion mobility mass spectrometry. Journal of the American Chemical Society. 2016;**138**:11254-11263. DOI: 10.1021/jacs.6b06205

[49] Amsharov KY. Towards rational synthesis of non-IPR fullerenes via tandem cyclode-hydrofluorination of fluoroarene-based precursors. Physica Status Solidi B: Basic Solid State Physics. 2017;**254**:1-5. DOI: 10.1002/pssb.201700170

Section 2

On Properties Aspects

Chapter 3

Reactivity Indexes and Structure of Fullerenes

Ernestina Mora Jiménez, Francisco J. Tenorio,
David Alejandro Hernández-Velázquez,
Jaime Gustavo Rodríguez-Zavala and
Gregorio Guzmán-Ramírez

Additional information is available at the end of the chapter

http://dx.doi.org/10.5772/intechopen.70642

Abstract

The discovery of fullerenes and their production in measurable quantities launched many studies about their reactivity and possible applications. Their peculiar structure opened possibilities for their study, initially replacing carbon atoms with alternative atoms. The surface also offers the possibility of attaching several species and the interior of their hollow structure represents a challenge because of the possibility of confining elements or molecules that may become less stable when attached to the exterior of the cage. These modifications may considerably affect both chemical and physical properties. In this chapter, we propose the encapsulation of 3–10 nitrogen atoms as aggregates inside the C_{70} cage. We also study the structures and reactivity indexes and the stabilization conferred as a result of being part of the fullerene. These aggregates are mainly of interest because of their possible application as energetic materials.

Keywords: polynitrogen, endohedral fullerenes, Density Functional Theory, reactivity indexes, C70, energetic materials

1. Introduction

Over the course of time, carbon materials have become important components not only everyday aspects of life, but also vanguard research. The enormous diversity concerning their uses and applications is constantly increasing, representing an area of constant development. Whether as part of a compound or in pure form, carbon has always attracted attention. At least in its purest forms, its structural diversity is extremely attractive (**Figure 1**).

Research on the formation and arrangement of carbon compounds that included long chains of this element in interstellar space lead to the discovery of fullerenes in 1985. Subsequently,

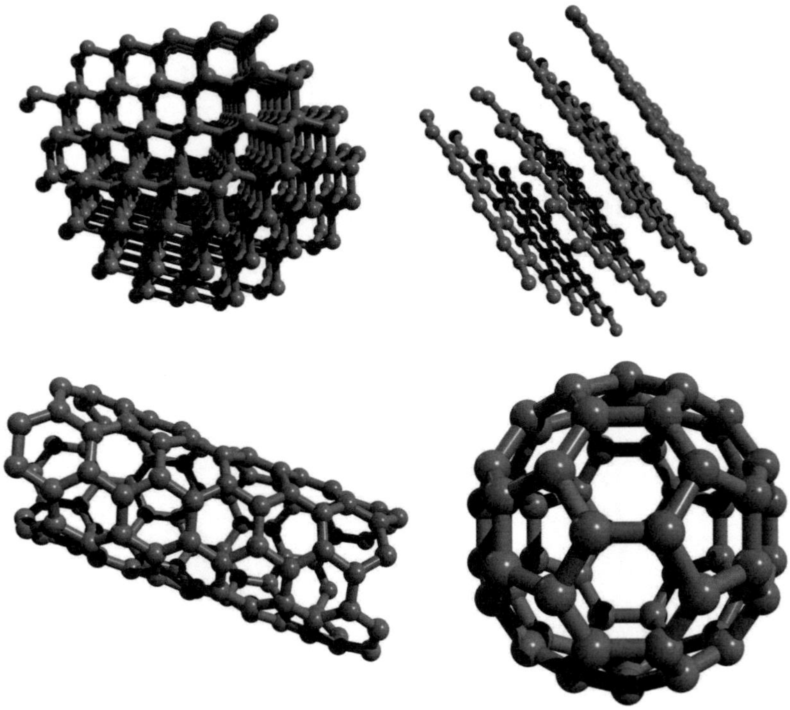

Figure 1. Allotropic forms of carbon. From left to right and up to down, diamond, graphite, nanotube and fullerene.

they were produced in measurable quantities for use in experiments [1–3]. Even though their study may have evolved accidentally, insufficient effort has been dedicated to understanding their structure, properties and applications. The best known examples of fullerenes are C_{60} and C_{70}, both of which consist of 12 pentagons, and only differ concerning the number of hexagons in their structure: 20 in the case of the first and 25 in the second. These differences result in the C_{60} structure, which is more like a soccer ball and C_{70}, which is more like a rugby ball (**Figure 2**). While C_{60} is more often known as a buckyball, the rest of the hollow structures are inscribed to the general fullerene group.

1.1. Types of fullerenes

Research into fullerenes has been extremely diverse, but can be categorized according to the structural aspect to be studied. Correspondingly, we have the following: exohedral fullerenes (with external ligands or molecules), endohedrals (with internal atoms or molecules) and heterofullerenes (with one or more carbon atoms in the cage replaced by heteroatoms).

Exohedral fullerenes have essentially resulted from efforts to "decorate" the surface of the fullerenes; mainly achieved by means of chemical functionalization. These types of fullerenes have proved to be extremely interesting, both for medical applications and for their possible applications in material sciences [4–8]. With respect to heterofullerenes, the substitution of carbon atoms by boron, nitrogen and silicon has been proposed; in the case of the first two, these have regularly

http://dx.doi.org/10.5772/intechopen.70642

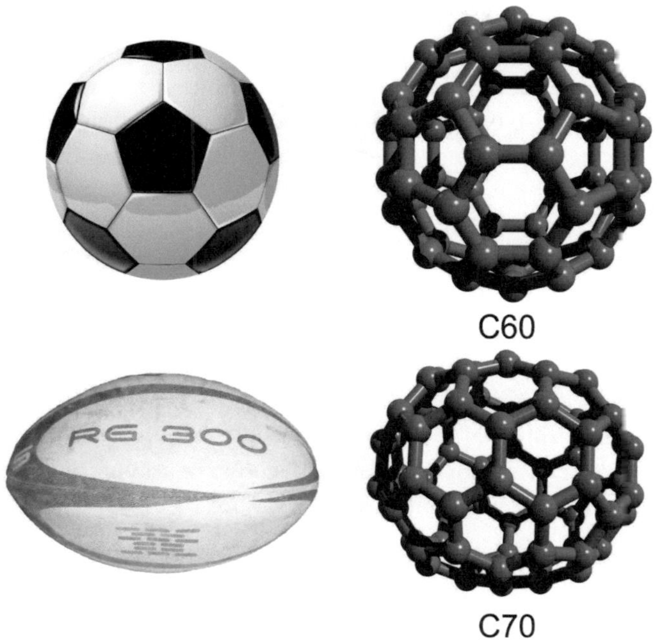

Figure 2. Comparison of C_{60} fullerene with a soccer ball and C_{70} with a rugby ball.

been as -BN- pairs, thus ensuring that isoelectronic species are obtained, although this has not necessarily been the only purpose; in the case of silicon, the main motive has been to search for useful applications in electronics [9–21]. Last but not least, endohedral fullerenes exist because of the fact that they are hollow; so they have the potential application of confining or protecting molecules, thus acting as carriers or stabilizers. Particularly, those with metallic atoms inside them—endohedral metallofullerenes—have been attractive for chemistry, physics and interdisciplinary areas, such as materials and biological sciences [22–25].

1.2. Polynitrogenated materials

Considerable interest in nitrogen compounds has emerged, especially those with a high content of this element, because of their particular properties and forms. Apart from the previously known molecular nitrogen (N_2), little was known of species with higher nitrogen content, until the synthesis of the azide anion (N_3^-) [26]. However, over time, new polynitrogenated species such as N_3, N_4, N_5^- and N_5^+ have been discovered or posited, whose synthesis has either represented a challenge or turned out to be surprisingly simple [27–30]. In spite of this, until now, the species with the highest content of nitrogen produced in measurable quantities has contained 5 nitrogen atoms, meaning that in particular the cyclopentazole anion could be considered as a motivator for theoretical studies on its formation, as well as representing a building block towards more complex structures [31–35]. As species with higher nitrogen content have proved to be very unstable with respect to decomposition to N_2, theoretical studies have outnumbered experimental ones. Of these, it has been determined that the preferred forms of N_4 and N_6 are acyclic, N_8 must be formed from an azide-pentazole (N_3–N_5) and that more complex polynitrogens may exist such as N_7, N_{12}, N_{18}, N_{20} and a great etcetera [36–42]. Considering larger structures, there is no doubt that polyhedral or fullerene-type structures have provided inspiration, as they are considered as alternative polynitrogenous arrangements with a greater number of nitrogen atoms [18, 19, 21, 41, 43–47]. In this sense, a hollow structure could in principle be useful for storing internal molecules, which is why it is common to find hollow carbon structures as candidates for confining polynitrogenous species. Currently, the most notable examples have been studies of $N@C_{60}$, $N@C_{70}$ or even $N_2@C_{60}$ [48–52], which have extended the use of the C_{60} cage as a way of confining up to 16 nitrogen atoms [53, 54]. Also recently, the encapsulation of a polymer nitrogen chain in a carbon nanotube was conceived, as in principle this could be stable up to room temperature [55]. This contrasts with other proposed polynitrogenated phases of nitrogen that have been inspired by the analogy to phosphorus and arsenic, all superior in energy to the "cubic gauche" (CG). These have been proposed for extreme conditions and some experimental evidence has even been found for them [56–61]. Therefore, confinement represents an alternative for the stabilization of polynitrogenated species that are not necessarily related to high temperatures and pressure.

As our work group has studied fullerenes and their reactivity [7, 16, 62–65] and it has been proposed that C_{60} might be an ideal candidate for trapping nitrogen by a polymer [53, 54], the intention now is to show that C_{70} might also be an alternative way of storing polynitrogenous species, contributing to the development of new materials in the future.

2. Methodology

Initiating with considerable structural diversity, we calculated and modeled endohedral fullerenes; $N_n@C_{70}$ (n = 3–10). For this, we used the C_{70} structure that corresponds to the isomer from this composition and complies with the isolated pentagon rule, IPR. The Avogadro visualizer was used as an auxiliary [66]. Given the considerable number of possible isomers, we determined the minimum energy structures in two stages. Initially, we obtained the geometry optimizations for all molecules, using the PM6 method [67]. To ensure that the global minimum for each composition has been identified, the use of search algorithms, such as those inspired by genetic algorithms, is essential, but this is currently beyond the scope of this type of system. However, the diversity of proven structures inspires confidence in our determination of the most important and representative species. Subsequently, both the geometry and the electronic structure were refined for the lower energy isomers of each composition. Likewise, the calculation of vibrational frequencies was undertaken in order to corroborate that the stationary points located on the potential energy surface correspond to a minimum (NImag = 0). All this was undertaken using the hybrid functional B3LYP with the base set 6-311G [68, 69]. As the structures of the polynitrogen species in free state may differ according to charge, this factor was also evaluated, determining the structures for the isomers neutral, cation and anion. All calculations were performed using the Gaussian09 program [70].

We also calculated ionization energies (IE) and electron affinities (EA). These were therefore calculated to reveal the following energy differences: $IE = E_{cation} - E_{neutral}$, $EA = E_{neutral} - E_{anion}$.

Chemical potential was computed by conceptual Density Functional Theory approximation. For an N electron system with an external potential v(r) and total energy E, electronegativity is defined as the partial energy derivative to the number of electrons at constant potential and then by the definition of Mulliken as the mean of IE and EA and the negative of electronegativity is the molecular chemical potential, μ:

$$\mu = -\chi = \left(\frac{\partial E}{\partial N}\right)_{v(r)} \approx -\frac{IE + EA}{2} \quad (1)$$

Chemical hardness was calculated as defined by Parr and Pearson [71], differentiating the chemical potential to the number of electrons, also at constant energy potential:

$$\eta = \left(\frac{\partial^2 E}{\partial N^2}\right)_{v(r)} \approx \frac{IE - EA}{2} \quad (2)$$

Energies were also obtained for the stabilization reactions for polymer species within the C_{70} structure, applying the formula: $\Delta E = \Sigma E_{(products)} - \Sigma E_{(reagents)}$ for two possible schemes. In the first, the stabilization of isolated nitrogen atoms could be analyzed through the reaction $n\mathrm{N} + C_{70} \rightarrow N_n@C_{70}$. In the second scheme, the reagents are substituted with nitrogen as found in standard state (N_2) or detected in experiments (N_3^-, N_5^-, etc).

3. Results

It was previously reported that the methodology consisted of an optimization at the PM6 level. As the refinement and electronic structure were performed at the B3LYP/6-311G level, they are presented exclusively in the following.

For the majority of the structures obtained, neutral and charged species correspond to the same arrangement. In cases where it is not observed, it will be mentioned in due course.

3.1. N*n*@C_{70} structures and reaction energies

N_3. The minimum energy structure corresponds to a linear and centered N_3, which shows the same structure in charged and neutral systems, see **Figure 3**. There is also found a second stable structure that may closely resemble the association between the N_2 molecule and an "isolated" nitrogen atom. The difference in the relative energies exceeds 40 kcal/mol, leading us to assume that the linear isomer encapsulated in C_{70} would predominate, if it was ever produced. Distances between nitrogen atoms (1.20 Å) suggest a weaker bond than that observed for N_2 (1.11 Å).

Reaction energies were calculated based on various assumptions. First, it is considering only the stabilization due to the encapsulation of nitrogen atoms as reagents. Second, considering structures that can be found at room temperature as starting materials (**Table 1**).

The $3N + C_{70} \rightarrow N_3@C_{70}$ reaction was used to evaluate the role of the C_{70} box for the stabilization (or not) of 3 isolated nitrogen atoms, compared to placing them within the cage in polymer

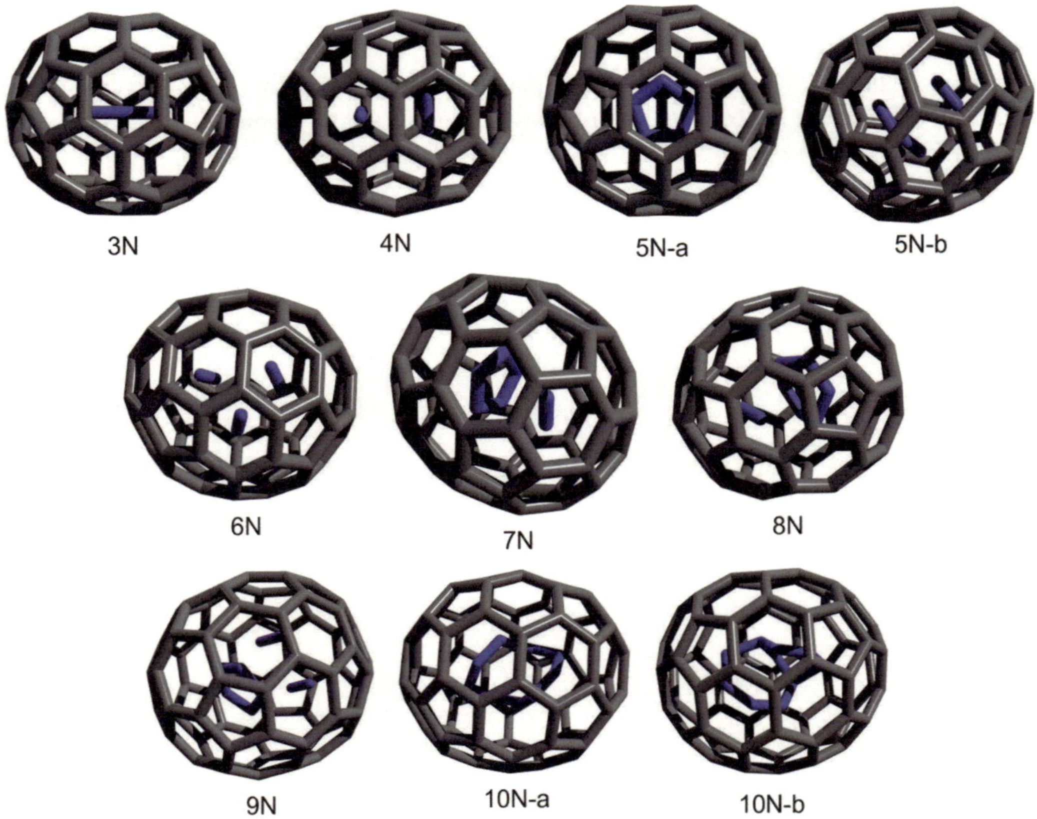

Figure 3. Structure of $N_n@C_{70}$, where $n = 3–10$, endohedral fullerenes, calculated using the B3LYP/6-311G method.

form as N_3. Apparently reactions are energetically favored. This energy difference is also highest for the cation, followed by the anion and finally the neutral (**Table 1**).

We evaluated energy differences for the reactions presented, while attempting to evaluate the role of the box, when initial material consisted of existing structures (such as N_3^- and N_2, rather than isolated nitrogen atoms). Evidently, as is stated in **Table 1**, the cation has lower ΔE energy. This molecule would be the most viable option at the time of the reaction, as the carbon cage would effectively stabilize the nitrogen polymer. Similarly, fullerene C_{70} is able to stabilize and cage the anion polynitrogen, as it also has a negative ΔE value. Contrastingly, in the case of the neutral structure, this is not predicted as a favorable reaction. In spite of this, stabilization of N_2 + N in the C_{70} cage is favorable (ΔE = −60.44 kcal/mol).

N_4. We analyzed structures consisting of two pairs of N_2 molecules, **Figure 3**, in almost parallel positions (with an approximate distance of 2.67 Å between the two N_2 pairs). An arrangement very similar to the neutral one was observed for the charged systems. Unstable structures for charged systems were also found, as in essence they could almost be considered as 4 atoms with no obvious interacting link.

The stabilization of arrangements of 4 nitrogen atoms in the cage was evaluated in terms of reaction energy, results for which are shown in **Table 1**. Evidently, the encapsulation would be

$Nn@C_{70}$	Anion	Neutral	Cation
For $n = 3$	$2N + N^- + C_{70} \rightarrow N_3@C_{70}^-$	$3N + C_{70} \rightarrow N_3@C_{70}$	$2N + N^+ + C_{70} \rightarrow N_3@C_{70}^+$
	−495.82	−383.89	−544.51
	$N_3^- + C_{70} \rightarrow N_3@C_{70}^-$	$1.5N_2 + C_{70} \rightarrow N_3@C_{70}$	$N_3^+ + C_{70} \rightarrow N_3@C_{70}^+$
	−39.13	101.29	−106.97
For $n = 4$	$3N + N^- + C_{70} \rightarrow N_4@C_{70}^-$	$4N + C_{70} \rightarrow N_4@C_{70}$	$3N + N^+ + C_{70} \rightarrow N_4@C_{70}^+$
	−678.69	−606.42	−764.55
	$N_2 + N_2^- + C_{70} \rightarrow N_4@C_{70}^-$	$2N_2 + C_{70} \rightarrow N_4@C_{70}$	$N_2 + N_2^+ + C_{70} \rightarrow N_4@C_{70}^+$
	−69.85	40.49	−145.73
For $n = 5$	$4N + N^- + C_{70} \rightarrow N_5@C_{70}^-$	$5N + C_{70} \rightarrow N_5@C_{70}$	$4N + N^+ + C_{70} \rightarrow N_5@C_{70}^+$
5 N-a	−764.51	−652.59	−814.12
5 N-b	−745.26	−646.04	−804.31
5 N-a	$N_5^- + C_{70} \rightarrow N_5@C_{70}^-$	$2.5N_2 + C_{70} \rightarrow N_5@C_{70}$	$2N_2 + N^+ + C_{70} \rightarrow N_5@C_{70}^+$
	−0.13	156.04	−167.21
5 N-b	$N_2 + N_3^- + C_{70} \rightarrow N_5@C_{70}^-$	$N_2 + N_3 + C_{70} \rightarrow N_5@C_{70}$	$N_2 + N_3^+ + C_{70} \rightarrow N_5@C_{70}^+$
	34.89	73.75	−43.31
For $n = 6$	$5N + N^- + C_{70} \rightarrow N_6@C_{70}^-$	$6N + C_{70} \rightarrow N_6@C_{70}$	$5N + N^+ + C_{70} \rightarrow N_6@C_{70}^+$
	−881.66	−807.24	−965.47
	$2N_2 + N_2^- + C_{70} \rightarrow N_6@C_{70}^-$	$3N_2 + C_{70} \rightarrow N_6@C_{70}$	$2N_2 + N_2^+ + C_{70} \rightarrow N_6@C_{70}^+$
	50.62	163.12	−23.19
	$N_5^- + N + C_{70} \rightarrow N_6@C_{70}^-$	$N_5^- + N^+ + C_{70} \rightarrow N_6@C_{70}$	$N_3 + N_3^+ + C_{70} \rightarrow N_6@C_{70}^+$
	−117.29	−385.00	−131.54
		$N_3^- + N_3^+ + C_{70} \rightarrow N_6@C_{70}$	
		−255.14	

Table 1. Reaction energies due to the encapsulation of nitrogen structures for $Nn@C_{70}$, where n = 3–6, obtained at B3LYP/6-311G level of theory.

energetically favored. Similar to that observed in the previous isomer, it is predicted that the interaction with the cationic system would be greater than with the anionic or with the neutral one.

Analyzing encapsulation reactions initiating with different materials (such as N_2, neutral and charged), evidently the neutral one is not predicted to be energetically favorable. This would lead to a structure which could be very interesting for possible high energy density material (HEDM) applications, a field where metastable compounds and single nitrogen-nitrogen bonds are encouraged. However, the distance between nitrogen atoms (1.11 Å) is characteristic for a triple bond, which discards (at least in principle) this structure as a candidate. For charged systems, the reaction energy is favorable. This gives an idea of the role played by the C_{70} cage in general, as it stabilizes systems with higher charge density.

N_5. The geometries for $N_5@C_{70}$ nitrogen polymers are shown in **Figure 3**. Captions 5 N-a and 5 N-b symbolize the optimized structures, noting that 5 N-a represents a cycle of 5 nitrogen atoms that is quite similar to cyclopentazole; a system found only as an anion [29, 31]. Distances between nitrogen atoms (~1.34 Å) for the cycle suggest a double-like bond. Contrastingly, 5 N-b isomers essentially consist of two systems; N_2 and N_3, separated by approximately 2.4 Å. We attempted to optimize a structure similar to N_5^+ that was predicted theoretically; however, our efforts were in vain, as all the structures presented fragmentation. Distances between nitrogen atoms are, for bended N_3 fragment, 1.19 Å, whereas for N_2 dimer 1.11 Å.

Geometry acquired by 5 N-a molecule represents the most stable. The energy of neutral 5 N-b is 6.5 kcal/mol greater than 5 N-a, which may be considerable, but would not be as difficult to achieve under experimental conditions. Likewise, the energy difference between the anion and cation systems exceeds 10 kcal/mol, with the cyclic isomers proving the most stable.

The formation energies of the different $N_5@C_{70}$ isomers were calculated by assuming that nitrogen atoms were the material at initiation, indicating at all times that this reaction would be favored. The highest value corresponded to cationic species, followed by anionic species and finally neutral species. Changing the reagents for other species reveals interesting dilemmas. Assuming formation with the pentazolate anion would imply a slightly favored reaction, but when substitute for $N_2 + N_3^-$, these systems become energetically unfavorable. Although this approach might discourage synthesis, it could also act as a motivator, as it would be considered as metastable and a candidate for high energy material, HEDMs.

N_6. For $N_6@C_{70}$, the most favored structure corresponds to the arrangement of three N_2 molecules, separated 2.4 Å each of the other (**Figure 3**). For previous reports of $N_6@C_{60}$ [53, 54], the lower energy structure corresponds to the pot-shaped hexagon, in contrast to that presented here. We suggest that this difference is due to the larger volume of C_{70} as compared to C_{60}, which allows a bigger dispersion within the cage. Cyclic structure with boat-shaped conformation is found 100 kcal/mol greater in energy than reported here and, therefore, is not considered as a possible candidate.

When the reactants are represented by isolated nitrogen atoms, formation energy shows an energetically favorable reaction. However, by changing the reactants for N_2, the prediction becomes highly unfavorable. We undertook analysis by changing the reagents, finding that when we proposed charged species as reactants, negative formation energies appeared. We attribute this mainly to the stabilization of charges by the cage.

N_7. Of isomers identified in this system, lowest energy consisted of an N_5–N_2 partnership, similar to that found previously for C_{60} [53, 54]. Other systems involving N_3 with $2N_2$ or even rings of N_7 proved to be stable but considerably higher in energy (**Figure 3**). Bond distances for nitrogen-nitrogen in the pentagon are 1.31–1.33 Å long, which indicate a double bond. The second energy isomer in order (with N_3 and $2N_2$ fragments) is 28 kcal/mol higher in energy, whereas a distorted heptagon N_7 cycle is about 90 kcal/mol higher. This is an indication that the pentagon-N_2 association would be the only isomer found in a hypothetical experiment.

Reaction energies show that the stabilization of individual nitrogen atoms is favored, whereas if the reagents are exchanged for others found experimentally such as $N_5^- + N_2$, this would not

http://dx.doi.org/10.5772/intechopen.70642

be the case. Again this presents a possible other use (as HEDM) for this type of molecule. We also present alternative reactions with energetically favored schemes, which we essentially attribute to the stabilization of charges in the fullerene cage (**Table 2**).

N_8. The most stable isomer for this composition consists of the assembly between a hexagon and a dimer as two separate units. Bond distances for N—N are around 1.33–1.34 Å, more related to a double bond. Likewise, isomers that involve the formation of a distorted—but unstable—octagon are found with considerably higher energies, making it possible to ensure

N*n*@C$_{70}$	**Anion (–)**	**Neutral (0)**	**Cation (+)**
For $n = 7$	$6N + N^- + C_{70} \rightarrow N_7@C_{70}^-$	$7N + C_{70} \rightarrow N_7@C_{70}$	$6N + N^+ + C_{70} \rightarrow N_7@C_{70}^+$
	−979.94	−866.37	−1026.35
	$N_2 + N_5^- + C_{70} \rightarrow N_7@C_{70}^-$	$3.5N_2 + C_{70} \rightarrow N_7@C_{70}$	$3N_2 + N^+ + C_{70} \rightarrow N_7@C_{70}^+$
	107.89	265.72	−55.99
	$2N_2 + N_3^- + C_{70} \rightarrow N_7@C_{70}^-$	$N_5^- + N_2^+ + C_{70} \rightarrow N_7@C_{70}$	
	123.67	−148.76	
For $n = 8$	$7N + N^- + C_{70} \rightarrow N_8@C_{70}^-$	$8N + C_{70} \rightarrow N_8@C_{70}$	$7N + N^- + C_{70} \rightarrow N_8@C_{70}^+$
	−1615.18	−1528.77	−1688.71
	$N_3 + N_3^- + N_2 \rightarrow N_8@C_{70}^-$	$4N_2 + C_{70} \rightarrow N_8@C_{70}$	$3.5N_2 + N^+ + C_{70} \rightarrow N_8@C_{70}^+$
	187.33	392.54	70.88
	$N_3 + N_5^- \rightarrow N_8@C_{70}^-$	$N_5^- + N_3^+ + C_{70} \rightarrow N_8@C_{70}$	
	171.55	−41.48	
For $n = 9$	$8N + N^- + C_{70} \rightarrow N_9@C_{70}^-$	$9N + C_{70} \rightarrow N_9@C_{70}$	$8N + N^+ + C_{70} \rightarrow N_9@C_{70}^+$
	−1046.85	−995.79	−1149.58
	$N_5^- + 2N_2 + C_{70} \rightarrow N_9@C_{70}^-$	$4.5N_2 + C_{70} \rightarrow N_9@C_{70}$	$4N_2 + N^+ + C_{70} \rightarrow N_9@C_{70}^+$
	364.43	459.75	144.23
	$3N_2 + N_3^- + C_{70} \rightarrow N_9@C_{70}^-$	$N_3^- + N_3^+ + N_3 + C_{70} \rightarrow N_9@C_{70}$	
	380.20	−47.35	
For $n = 10$	$9N + N^- + C_{70} \rightarrow N_{10}@C_{70}^-$	$10N + C_{70} \rightarrow N_{10}@C_{70}$	$9N + N^+ + C_{70} \rightarrow N_{10}@C_{70}^+$
b	−1143.41	−981.87	−1192.45
a	−1150.42	−996.12	−1184.66
	$N_3^- + 3.5N_2 + C_{70} \rightarrow N_{10}@C_{70}^-$	$5N_2 + C_{70} \rightarrow N_{10}@C_{70}$	$4N_2 + N_2^+ + C_{70} \rightarrow N_{10}@C_{70}^+$
a	438.36	621.14	396.74
	$N_5^- + 2.5N_2 \rightarrow N_{10}@C_{70}^-$		
	422.59		

Table 2. Reaction energies due to the encapsulation of nitrogen structures for N*n*@C$_{70}$, where n = 7–10, obtained at B3LYP/6-311G level of theory.

that this isomer would predominate in an experiment. The nitrogen hexagon shows distances that suggest double bonds, ranging from 1.31 to 1.35 Å, whereas for the unstable octagon they range from 1.26 to 1.39 Å.

Similarly, the reaction energy from the stabilization of nitrogen atoms shows that this would be a favored case, starting with the cation, followed by the anion and finally the neutral. Changing the reactants it is found that metastable species would be formed with molecules present at laboratory conditions, such as N_2 and N_5^-. This alternate reaction scheme opens up the possibility to produce HEDM candidates.

N_9. The most favored structure for this encapsulation consists of the assembly formed from an N_5 ring and two dimers parallel to this. Nitrogen-nitrogen bonds for the cycle are very similar to the corresponding for similar cycles (1.3–1.33 Å) and the distance to N_2 is about 2.29 Å. Qualitatively, it is similar to that found for C_{60}, although with considerably greater distances [53, 54]. Stable structures involving ensembles formed with N_3 and $3N_2$ species are 18.78 kcal/mol higher in energy. The fact that it is preferred a more complex structure instead simple ones, indicates us to suggest a more determinant role of the C_{70} cage in the stabilization of the structure. The negative values of the reaction energy reflect the stabilization of the polynitrogenated species and evidently the reaction is more favored when the system is positively charged, after which follows the negative system and finally the neutral system. The energy difference revealing a positive energy reflects an unfavorable and endergonic reaction, although probably with metastable products. Moreover, by exploring other reaction paths to obtain the isomers of 9 nitrogen atoms, favorable reactions are obtained.

N_{10}. As the amount of nitrogen atoms in the fullerene cage increases, a greater variety of nitrogen polymer structures can be perceived. For this system, the neutral minimum corresponds to a distorted cycle of 10 atoms, which contrasts with that reported previously for C_{60} [53, 54]. However, the second energy structure for charged systems (with an energy difference not exceeding 8 kcal/mol compared to the most stable) corresponds to that reported for C_{60}. Bonding distances between nitrogen atoms are 1.23–1.43 Å for the lowest energy isomer and 1.24–1.48 Å for the second. The formation energy starting from isolated nitrogen atoms shows negative values, suggesting stabilization on the part of the box for N_{10}, which is highest for the cation, followed by the anion and finally the neutral. The formation reaction that initiates from a variety of reagents, such as N_3^-, N_5^- and N_2, shows positive values, which suggest metastable structures.

3.2. Global reactivity indexes

We obtained the following chemical reactivity indexes: ionization potential, electron affinity, as well as electronic hardness and chemical potential and the results are summarized in **Table 3**. For ionization potentials, apparently fullerenes encapsulating 3–8 nitrogen atoms fall within a range of 7.4–7.6 eV. Notably, when the fullerenes contain 4 and 6 nitrogen atoms, the force with which the nitrogen endohedral fullerenes retain the electrons is slightly greater than when encapsulating between 3 and 5 nitrogen atoms. When encapsulating between 7 and 8 nitrogen atoms, this force tends to be the same, but with 9 nitrogen atoms within the

http://dx.doi.org/10.5772/intechopen.70642

	I	A	μ	η
$N_3@C_{70}$	7.51	4.49	−6	1.51
$N_4@C_{70}$	7.61	2.77	−5.19	2.42
$N_5@C_{70}$	7.47	4.49	−5.98	1.49
$N_6@C_{70}$	7.61	2.87	−5.24	2.37
$N_7@C_{70}$	7.54	4.56	−6.05	1.49
$N_8@C_{70}$	7.54	3.39	−5.46	2.08
$N_9@C_{70}$	7.81	1.85	−4.83	2.98
$N_{10}@C_{70}$	6.3	6.33	−6.31	−0.02

Table 3. Global reactivity indexes for N*n*@C_{70}, where *n* = 7–10, obtained at B3LYP/6-311G level of theory.

fullerene, this energy increases to 7.81 eV, implying that more energy is required in order to remove an electron. Contrastingly, when there are 10 nitrogen atoms within the fullerene network, it is easier to remove an electron because the ionization potential decreases. The even/odd effect observed in the first clusters may be explained by the resistance of systems with closed shell (even number of nitrogens) to be ionized, in contrast to those with an open shell (odd number of nitrogens) that contrarily would present a greater tendency to lose an electron, resulting in a closed shell.

The odd/even behavior of endohedral fullerenes, on gaining an electron and forming a negative ion, is also reflected in electron affinity; as apparently when C_{70} fullerenes encapsulate 3, 5 and 7 nitrogen atoms, the energy released is greater than when 4, 6 and 8 nitrogen atoms are caged. Notably, the energy of 8 nitrogen atoms increases to a lesser degree than 4 and 6 atoms of nitrogen. When encapsulating 9 and 10 nitrogen atoms, the sequence tends to change, because higher energy release corresponds to the 10 nitrogen structure and the lower energy release corresponds to the 9 nitrogen structure.

In terms of electronic chemical potential, endohedral fullerenes: of 4, 6 and 8 N atoms present the highest values for chemical potential, so they tend to accept electrons. Fullerenes of 3, 5 and 7 atoms have the lowest value, so they can donate electrons more easily. However, on reaching 9 and 10 atoms, the sequence changes because the isomer with 9 nitrogen atoms is the one that has the highest value of chemical potential than all other values, whereas the one with 10 nitrogen atoms has the lowest value of all.

With respect to overall hardness, there is a sequence ranging from 3 to 7 nitrogen atoms where the highest hardness values are the even numbers. These, therefore, have the greatest resistance for modifying their electronic density. The lowest hardness values correspond to the odd numbers of nitrogen atoms. At 8 nitrogen atoms, the sequence changes direction: now the odd number of nitrogen atoms is the one that presents the greatest value for global hardness, besides being the molecule that is least reactive in terms of hardness. Considering the case of N_{10}, this structure has the lowest hardness value. Consequently, it can modify

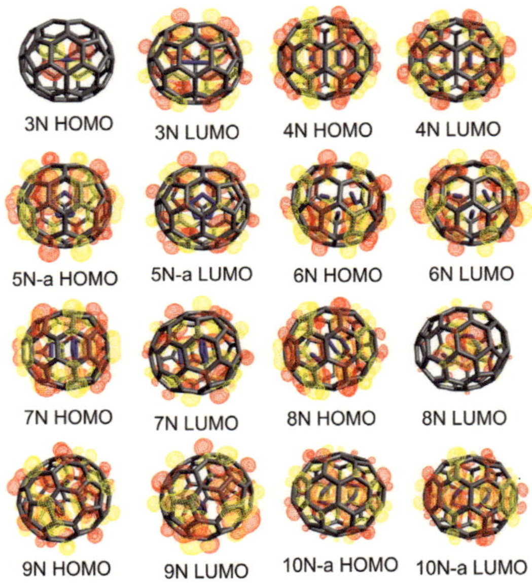

Figure 4. Frontier orbitals for $N_n@C_{70}$, where $n = 3–10$, endohedral fullerenes, calculated using the B3LYP/6-311G method.

its electronic density more easily, so that it represents the most reactive molecule. Previous results for C_{60} encapsulation [53] showed a reduced HOMO-LUMO gap, as the number of nitrogens increased: 2, 4, 6, 8, 10. This is somewhat similar to the results for hardness, as one of the approximations to this parameter is precisely the HOMO-LUMO gap (**Figure 4**).

3.3. Analysis of frontier orbitals

Previous studies have mentioned that from 2 to 10 nitrogen atoms, there would be a charge transfer from the C_{60} cage to the nitrogen polymer, explained as resulting from the greater ionization and electronegativity potential of N compared to C. For more than 10 nitrogen atoms, there would be a reverse transfer from the nitrogen to the carbon atoms. This was attributed to reduced space availability and overlapping of orbitals [53]. In this work, the frontier orbitals of the most stable structures were analyzed, showing that there is an appreciable contribution from frontier orbitals to the nitrogen atoms. This implies that there must be areas susceptible to receiving charge density. It is especially interesting that when the number of N atoms within the fullerene increases, the contribution of Nitrogen-centered orbitals for LUMO is even greater. Particularly, in the case of N_8 and higher systems, a very considerable contribution of HOMO is observed, and a sure indication that possible interactions with the cage are initiating. This may also relate to the suggestion that greater interaction between the cage and the N_n takes place, as the amount of nitrogen atoms increases. The reaction energy may also result from this, particularly as it is apparent that from N_3 to N_7 there is a more or less constant increase of 150 kcal/mol in stabilization energy. This reaction energy changes with N_8 and higher, suggesting greater participation on the part of the cage and its interaction with N_n.

4. Conclusions

The C_{70} fullerene is viable for storage and stabilization of nitrogen aggregates of at least 3–10 atoms, without presenting any concurrent structural deformation in the carbon network or obvious interaction (bond-like) between the nitrogen atoms and the carbon of the fullerene.

Evidently, the formation energy from the isolated nitrogen atoms is favored in all cases and from N_3 to N_8 there is a progressive increase, indicating the role of the cage as an encapsulator and stabilizer of polynitrogenous structures. When we propose the formation of endohedral fullerenes, initiating with materials that can be found under laboratory conditions, we predict that metastable structures will form, a fact that may be of interest in the potential use of these materials as HEDMs. The stabilizing contribution of the cage becomes more evident, as the number of nitrogen atoms within it increases, manifested in a considerable contribution on the part of frontier orbitals as potential charge stabilizers.

As C_{70} fullerene is a large molecule of approximately 1 nm diameter, nitrogen, polymers of 3–7 nitrogen atoms prefer to remain as nitrogen molecules or as azides. However, from 8 atoms onwards, the fullerene cage begins to have greater interaction with the nitrogen polymers, reflected in the fact that these begin to compact and create more complex nitrogen polymers (in the form of rings).

According to indexes of global chemical reactivity, endohedral fullerenes present an odd/even behavior that corresponds to that expected for open layer species (odd number of electrons: lower ionization potential and hardness, greater electronic affinity and chemical potential) and closed layer (even number of electrons: greater ionization potential and hardness, lower electronic affinity and chemical potential). However, for 8, 9 and 10 encapsulated nitrogen atoms, there is a change in behavior. This coincides with the change in formation energies and the analysis of frontier orbitals, reflecting greater participation of the cage in fullerene behavior. The fact that smaller fullerenes have been able to encapsulate more atoms is an indication that the point of saturation in this structure has not yet been reached, providing an incentive to find new structures with greater complexity, as the degree of encapsulation progresses. These would thus constitute candidates for energetic materials and studies are ongoing.

Acknowledgements

The authors like to thank PRODEP (formerly PROMEP) for support provided through the 103.5/13/6900 office. FJTR like to thank the University of Guadalajara for authorizing sabbatical leave. DAHV like to thank CONACYT for support provided by the program "Apoyos para la Incorporación de Investigadores Vinculada a la Consolidación Institucional de Grupos de Investigación y/o Fortalecimiento del Posgrado Nacional." CONACYT through Project 52,827 is also acknowledged.

Author details

Ernestina Mora Jiménez[1], Francisco J. Tenorio[1,2]*, David Alejandro Hernández-Velázquez[1], Jaime Gustavo Rodríguez-Zavala[1] and Gregorio Guzmán-Ramírez[3]

*Address all correspondence to: ftenorio@culagos.udg.mx

1 Departamento de Ciencias Exactas y Tecnología, Centro Universitario de los Lagos, Universidad de Guadalajara, Lagos de Moreno, Jalisco, México

2 Facultad de Química, Departamento de Farmacia, Universidad Nacional Autónoma de México, Ciudad de México, México

3 Departamento de Ingenierías, Centro Universitario de Tonalá, Universidad de Guadalajara, Tonalá, Jalisco, México

References

[1] Kroto HW, Heath JR, O'Brien SC, Curl RF, Smalley RE. C60: Buckminsterfullerene. Nature. 1985 Nov 14;**318**(6042):162-163

[2] Kratschmer W, Lamb LD, Fostiropoulos K, Huffman DR. Solid C60: A new form of carbon. Nature. 1990 Sep 27;**347**(6291):354-358

[3] Wragg JL, Chamberlain JE, White HW, Kratschmer W, Huffman DR. Scanning tunnelling microscopy of solid C60/C70. Nature. 1990 Dec 13;**348**(6302):623-624

[4] Prato M. [60]Fullerene chemistry for materials science applications. Journal of Materials Chemistry. 1997;**7**(7):1097-1109

[5] Da Ros T, Prato M. Medicinal chemistry with fullerenes and fullerene derivatives. Chemical Communications. 1999;**8**:663-669

[6] Wudl F. Fullerene materials. Journal of Materials Chemistry. 2002;**12**(7):1959-1963

[7] Rodriguez-Zavala JG, Tenorio FJ, Samaniego C, Mendez-Barrientos CI, Pena-Lecona FG, Munoz-Maciel J, et al. Theoretical study on the sequential hydroxylation of C-82 fullerene based on Fukui function. Molecular Physics. 2011;**109**(14):1771-1783

[8] Yan W, Seifermann SM, Pierrat P, Bräse S. Synthesis of highly functionalized C_{60} fullerene derivatives and their applications in material and life sciences. Organic & Biomolecular Chemistry. 2015;**13**(1):25-54

[9] Oku T, Nishiwaki A, Narita I. Formation and atomic structures of BnNn (n = 24-60) clusters studied by mass spectrometry, high-resolution electron microscopy and molecular orbital calculations. Physica B: Condensed Matter. 2004 Aug 15;**351**(1-2):184-190

[10] Golberg D, Bando Y, Stéphan O, Kurashima K. Octahedral boron nitride fullerenes formed by electron beam irradiation. Applied Physics Letters. 1998 Oct 21;**73**(17):2441-2443

[11] Oku T, Narita I, Nishiwaki A. Formation and structures of B36N36 and Y@B36N36 clusters studied by high-resolution electron microscopy and mass spectrometry. Journal of Physics and Chemistry of Solids. 2004 Mar;**65**(2-3):369-372

[12] Blase X, Rubio A, Louie S, Cohen M. Stability and Band Gap Constancy of Boron Nitride Nanotubes. EPL (Europhysics Letters). 2007;**28**(5):335-340

[13] Nishiwaki A, Oku T, Suganuma K. Atomic and electronic structures of endohedral B36N36 clusters with doping elements studied by molecular orbital calculations. Physica B: Condensed Matter. 2004 Jun 15;**349**(1-4):254-259

[14] Harrison PG. Silicate cages: Precursors to new materials. Journal of Organometallic Chemistry. 1997;**542**(2):141-183

[15] Bernal Texca FG, Chigo-Anota E, Tepech Carrillo L, Castro M. A DFT study of the electronic and magnetic properties of C36Si24 fullerenes. Computational & Theoretical Chemistry. 2017 [cited 19 May 2017];**1103**:1-10

[16] Tenorio FJ, Robles J. On the stability and reactivity of C-Si heterofullerenes. International Journal of Quantum Chemistry. 2000 Oct;**80**(2):220-226

[17] Baei MT. Remove of toxic pyridine from environmental systems by using B12N12 nanocage. Superlattices and Microstructures. 2013;**58**:31-37

[18] Wang LJ, Zgierski MZ. Super-high energy-rich nitrogen cluster N60. Chemical Physics Letters. 2003;**376**:698-703

[19] Wright JS, McKay DJ, DiLabio GA. Dodecahedral molecular nitrogen (N20) and related structures. Journal of Molecular Structure: THEOCHEM. 1998;**424**(1):47-55

[20] Su B, Feng X, Guo X, Li N. Polynitrogen clusters encapsulated inside B24N24 fullerene-like nanocages: Nanoscale high energy materials studied by density functional theory. Inorganica Chim Acta. 2017;**456**:128-135

[21] Bliznyuk AA, Shen M, Schaefer HF. The dodecahedral N20 molecule. Some theoretical predictions. Chemical Physics Letters. 1992;**198**(3):249-252

[22] Shinohara H. Endohedral metallofullerenes. Reports Progress in Physics. 2000;**63**(6):843

[23] Cerón MR, Maffeis V, Stevenson S, Echegoyen L. Endohedral fullerenes: Synthesis, isolation, mono- and bis-functionalization. Inorganica Chim Acta. 2017. DOI:10.1016/j.ica.2017.03.040. In press

[24] Yeh W-Y. Coordination and reactivity of functionalized fullerenes, open-cage fullerenes, and endohedral metallofullerenes by organometallic complexes. Journal of Organometallic Chemistry. 2015;**784**:13-23

[25] Popov AA, Yang S, Dunsch L. Endohedral Fullerenes. Chemical Reviews. 2013 Aug 14;**113**(8):5989-6113

[26] Curtius T. Ueber Stickstoffwasserstoffsäure (Azoimid) N3H. Berichte der Dtsch Chem Gesellschaft. 1890 Jul 1;**23**(2):3023-3033

[27] Thrush BA. The detection of free radicals in the high intensity photolysis of hydrogen Azide. Proceedings of the Royal Society of London Series A Mathematical Physical Science. 1956 Apr 10;**235**(1200):143 LP-143147

[28] Cacace F, de Petris G, Troiani A. Experimental detection of Tetranitrogen. Science (80-). 2002 Jan 18;**295**(5554):-480 LP, 481

[29] Vij A, Pavlovich JG, Wilson WW, Vij V, Christe KO. Experimental detection of the Pentaazacyclopentadienide (Pentazolate) anion, cyclo-N5−. Angewandte Chemie International Edition. 2002 Aug 16;**41**(16):3051-3054

[30] Christe KO, Wilson WW, Sheehy JA, Boatz JA. N5+: A novel Homoleptic Polynitrogen ion as a high energy density material. Angewandte Chemie International Edition. 1999 Jul 12;**38**(13-14):2004-2009

[31] Bazanov B, Geiger U, Carmieli R, Grinstein D, Welner S, Haas Y. Detection of Cyclo-N5− in THF solution. Angewandte Chemie International Edition. 2016 Oct 10;**55**(42):13233-13235

[32] Zhang C, Yang C, Hu B, Yu C, Zheng Z, Sun C. A symmetric co$(N_5)_2(H_2O)_4$·4 H_2O high-nitrogen compound formed by cobalt(II) cation trapping of a cyclo-N_5− anion. Angewandte Chemie International Edition. 2017 Apr 10;**56**(16):4512-4514

[33] Geiger U, Haas Y. Preparation of the Cyclopentazole anion in the bulk: A computational study. The Journal of Physical Chemistry. B. 2016 Jul 7;**120**(26):6208-6214

[34] Choi C, Yoo H-W, Goh EM, Cho SG, Jung Y. Ti(N5)4 as a potential nitrogen-rich stable high-energy density material. The Journal of Physical Chemistry. A. 2016 Jun 23;**120**(24):4249-4255

[35] Cui J, Zhang Y, Zhao F, Yang J, Shen G, Xu Y. HB (N5)3M (M = Li, Na, K, and Rb): A new kind of pentazolides as HEDMs. Progress in Natural Science. 2009;**19**(1):41-45

[36] Wang X, Ren Y, Shuai M-B, Wong N-B, Li W-K, Tian A-M. Structure and stability of new N7 isomers. Journal of Molecular Structure: THEOCHEM. 2001;**538**(1):145-156

[37] Gagliard L, Evangelisti S, Roos BO, Widmark P-O. A theoretical study of ten N8 isomers. Journal of Molecular Structure: THEOCHEM. 1998;**428**(1):1-8

[38] Hirshberg B, Gerber RB. Decomposition mechanisms and dynamics of N6: Bond orders and partial charges along classical trajectories. Chemical Physics Letters. 2012;**531**:46-51

[39] Klapötke TM. Ab initio calculations of the open-chain N6 diazide molecule. Journal of Molecular Structure: THEOCHEM. 2000 Mar 17;**499**(1-3):99-104

[40] Klapötke TM, Harcourt RD. The interconversion of N12 to N8 and two equivalents of N2. Journal of Molecular Structure: THEOCHEM. 31 May 2001;**541**(1-3):237-242

[41] Gu JD, Chen KX, Jiang HL, Chen JZ, Ji RY, Ren Y, Tian AM. N18: A computational investigation. Journal of Molecular Structure: THEOCHEM. 1998 Mar 23;**428**(1-3):183-188

[42] Sturdivant SE, Nelson FA, Strout DL. Trends in stability for N18 cages. The Journal of Physical Chemistry. A. 2004 Aug 1;**108**(34):7087-7090

[43] Balasubramanian K. Nuclear spin multiplets for the dodecahedral N20 cluster. Chemical Physics Letters. 1993;**202**(3):271-277

[44] Ha T-K, Suleimenov O, Nguyen MT. A quantum chemical study of three isomers of N20. Chemical Physics Letters. 1999;**315**(5):327-334

[45] Liu F-L, Yang F, Zhang L-X. Computational note on C10N30: Two intact dodecahedral C5N15 cages linked by five carbon–carbon single bonds. Journal of Molecular Structure: THEOCHEM. 2010;**950**(1-3):98-99

[46] Zhou H, Beuve M, Yang F, Wong N-B, Li W-K. Theoretical investigation on the cylinder-shaped N66 cage. Computational & Theoretical Chemistry. 2013 Feb 1;**1005**:68-74

[47] Zhou H, Wong N-B, Tian A. Theoretical study on the cylinder-shaped N78 cage. Journal of Molecular Graphics & Modelling. 2006 Dec;**25**(4):578-583

[48] Knapp C, Dinse K-P, Pietzak B, Waiblinger M, Weidinger A. Fourier transform EPR study of N@C60 in solution. Chemical Physics Letters. 1997;**272**(5):433-437

[49] Dinse K-P. EPR investigation of atoms in chemical traps. Physical Chemistry Chemical Physics. 2002;**4**(22):5442-5447

[50] Meyer C, Harneit W, Lips K, Weidinger A, Jakes P, Dinse KP. Alignment of the Endohedral fullerenes N@C60 and N@C70 in a liquid-crystal matrix. Physical Review A. 2002;**65**:61201

[51] Dietel E, Hirsch A, Pietzak B, Waiblinger M, Lips K, Weidinger A, et al. Atomic nitrogen encapsulated in fullerenes: Effects of cage variations. Journal of the American Chemical Society. 1999;**121**:2432

[52] Franco L, Ceola S, Corvaja C, Bolzonella S, Harneit W, Maggini M. Synthesis and magnetic properties of N@C60 derivatives. Chemical Physics Letters. 2006;**422**(1):100-105

[53] Sharma H, Garg I, Dharamvir K, Jindal VK. Structure of polynitrogen clusters encapsulated in C60: A density functional study. Journal of Physical Chemistry C. 20 May 2010;**114**(19):9153-9160

[54] Sharma H, Garg I, Dharamvir K, Jindal VK. Structural, electronic, and vibrational properties of C60–nNn (n = 1–12). The Journal of Physical Chemistry. A. 2009 Aug 6; **113**(31):9002-9013

[55] Zhong Y, Jaidann M, Zhang Y, Zhang G, Liu H, Ioan Ionescu M, et al. Synthesis of high nitrogen doping of carbon nanotubes and modeling the stabilization of filled DAATO@ CNTs (10,10) for nanoenergetic materials. Journal of Physics and Chemistry of Solids. 2010;**71**(2):134-139

[56] Mitáš L, Martin RM. Quantum Monte Carlo of nitrogen: Atom, dimer, atomic, and molecular solids. Physical Review Letters. 1994 Apr 11;**72**(15):2438-2441

[57] Mailhiot C, Yang LH, McMahan AK. Polymeric nitrogen. Physical Review B. 1992;**46**(22): 14419-14435

[58] Lewis SP, Cohen ML. High-pressure atomic phases of solid nitrogen. Physical Review B. 1992;**46**(17):11117-11120

[59] Martin RM, Needs RJ. Theoretical study of the molecular-to-nonmolecular transformation of nitrogen at high pressures. Physical Review B. 1986;**34**(8):5082-5092

[60] Eremets MI, Gavriliuk AG, Trojan IA, Dzivenko DA, Boehler R. Single-bonded cubic form of nitrogen. Nature Materials. 2004;**3**(8):558-563

[61] Mattson WD, Sanchez-Portal D, Chiesa S, Martin RM. Prediction of New Phases of Nitrogen at High Pressure from First-Principles Simulations. Physical Review Letters. 2004;**93**(12):125501

[62] Rodríguez-Zavala JG, Barajas-Barraza RE, Padilla-Osuna I, Guirado-López RA. Hydration behaviour of polyhydroxylated fullerenes. Journal of Physics B: Atomic, Molecular and Optical Physics. 2011;**44**(20):205104

[63] Rodríguez-Zavala JG, Padilla-Osuna I, Guirado-López RA. Modifying the endohedral la-ion location in both neutral and positively charged polyhydroxylated metallofullerenes. Physical Review B. 2008 Oct 24;**78**(15):155426

[64] Rodríguez-Zavala JG, Guirado-López RA. Stability of highly OH-covered C60 fullerenes: Role of Coadsorbed O impurities and of the charge state of the cage in the formation of carbon-opened structures. The Journal of Physical Chemistry. A. 2006 Aug 1;**110**(30):9459-9468

[65] Rodríguez-Zavala JG, Guirado-López RA. Structure and energetics of polyhydroxylated carbon fullerenes. Physical Review B. 2004 Feb 23;**69**(7):75411

[66] Hanwell MD, Curtis DE, Lonie DC, Vandermeersch T, Zurek E, Hutchison GR. Avogadro: An advanced semantic chemical editor, visualization, and analysis platform. Journal of Cheminformatics. 2012;**4**(1):17

[67] Stewart JJP. Optimization of parameters for semiempirical methods V: Modification of NDDO approximations and application to 70 elements. Journal of Molecular Modeling. 2007;**13**(12):1173-1213

[68] Becke AD. A new mixing of Hartree–Fock and local density-functional theories. The Journal of Chemical Physics. 1993;**98**(2):1372-1377

[69] Krishnan R, Binkley JS, Seeger R, Pople JA. Self-consistent molecular orbital methods. XX. A basis set for correlated wave functions. The Journal of Chemical Physics. 1980 Jan 1;**72**(1):650-654

[70] Frisch MJ, Trucks GW, Schlegel HB, Scuseria GE, Robb MA, Cheeseman JR, Scalmani G, Barone V, Mennucci B, Petersson GA, Nakatsuji H, Caricato M, Li X, Hratchian HP, Izmaylov AF, Bloino J, Zheng G, Sonnenberg JL, Hada M, Ehara M, Toyota K, Fukuda R, Hasegawa J, Ishida M, Nakajima T, Honda Y, Kitao O, Nakai H, Vreven T, Montgomery JA Jr, Peralta JE, Ogliaro F, Bearpark M, Heyd JJ, Brothers E, Kudin KN, Staroverov VN, Kobayashi R, Normand J, Raghavachari K, Rendell A, Burant JC, Iyengar SS, Tomasi J,

Cossi M, Rega N, Millam JM, Klene M, Knox JE, Cross JB, Bakken V, Adamo C, Jaramillo J, Gomperts R, Stratmann RE, Yazyev O, Austin AJ, Cammi R, Pomelli C, Ochterski JW, Martin RL, Morokuma K, Zakrzewski VG, Voth GA, Salvador P, Dannenberg JJ, Dapprich S, Daniels AD, Farkas Ö, Foresman JB, Ortiz JV, Cioslowski J, Fox DJ. Gaussian, Inc., Wallingford CT, 2009. Gaussian 09, Revision C.01

[71] Parr RG, Pearson RG. Absolute hardness: Companion parameter to absolute electronegativity. Journal of the American Chemical Society. 1983 Dec 1;**105**(25):7512-7516

Chapter 4

Fullerene Dynamics with X-Ray Free-Electron Lasers

Nora Berrah

Additional information is available at the end of the chapter

http://dx.doi.org/10.5772/intechopen.70769

Abstract

Ultrafast and ultra-intense, short X-ray pulses from free-electron lasers (FELs) have opened up a new regime for all scientific research and for fullerenes in particular. FELs allow for the investigation of ultrafast nonlinear and multiphoton processes, as well as the exploration of the fragmentation dynamics of fullerenes. This chapter describes the FELs' attributes that enable new FEL-based investigations. In particular, we report on the X-ray ionization and fragmentation of C_{60} under high- and mid-fluence femtosecond pulses, from the Linac Coherent Light Source (LCLS), at SLAC National Accelerator Laboratory. We also describe the X-ray ionization and fragmentation with low-fluence X-ray pulses of the endohedral fullerene ($Ho_3N@C_{80}$). We end our contribution by presenting opportunities for future time-resolved dynamics research using pump-probe techniques.

Keywords: fullerenes, photoionization, fragmentation, ions, photons, free-electron laser, FEL, LCLS, femtosecond, attosecond, X-rays, dynamics, pump probe, C_{60}, $Ho_3N@C_{80}$

1. Introduction

Fullerenes, which bridge the gap between molecules and nanoparticles, are ideal systems for investigating the dynamical behavior of extended systems when exposed to X-rays. Fullerenes and their derivatives, characterized by their hollow geometric structures and nanometer-sized outer diameter, draw a great deal of interest due to their wide range of applications and "supramolecular" physical and chemical properties [1]. Fullerenes have displayed molecular [2] and bulk [3] behavior and have proven to be an excellent testing ground for experiments and theories [4].

This chapter, which is not a review, focuses on the interaction of intense lasers with fullerenes—and in particular, with X-ray free-electron lasers (FELs). After brief, general introductions on the interaction of fullerenes with strong laser fields using tabletop lasers and on endohedral fullerenes, ultrafast and ultra-intense free-electron lasers (FELs) are described; we focus our report on the interaction of C_{60} and of the endohedral $Ho_3N@C_{80}$ fullerene with a specific FEL: the Linac Coherent Light Source (LCLS) at SLAC National Accelerator Laboratory.

In recent years, the nonlinear physics research in atoms, molecules, and clusters that were conducted using strong laser fields has led to various phenomena, such as the generation of attosecond pulses [5]. The behavior of molecules in short, intense laser fields [6] was extended to large molecules, such as C_{60}, which is intriguing due to the numerous nuclei-electron responses exhibited, because it is a cage of 60 atoms with 240 valence-electrons [7–16]. The interaction of such a large system is key to investigating many-body problems induced on the system's electrons by the photon electric field. The photon interaction with the electronic fullerene's degrees of freedom results in electronic dynamics that lead to nuclei dynamics, because they are both interconnected. Fullerenes, including endohedral fullerenes, are ideal candidates to explore their many-body responses to electromagnetic fields because they respond in different ways—depending upon the field parameters [7–17]. Ionization, which is one of the possible reactions, has shown to occur on different time scales.

The laser photoionization mechanisms of fullerenes have been found to be wavelength and pulse duration-dependent [9, 10, 18]. For IR pulses (800 nm) of about 30 fs duration and intensities below 5×10^{13} W/cm^2, it was found that multiphoton processes dominate when ionizing C_{60}, while tunneling, and/or over-the-barrier ionization and ionization due to induced electron re-collision [8] have a low probability to occur under these conditions. The single-active-electron (SAE) method was used to calculate the ionization of C_{60} in intense, 4×10^{13} W/cm^2 laser pulses with durations between 27 and 70 fs, and for a wide range of wavelengths ranging from 395 to 1800 nm [19]. This calculation agreed with measurements by Shchatsinin et al. [12]. For a long IR wavelength of 1800 nm and 70 fs pulse duration, the SAE picture predicts "over the barrier" ionization for a peak intensity of 10^{15} W/cm^2, leading to non-fragmented but highly charged C_{60}^{q+} (q = 1–12) [7]. At a short wavelength of 355 nm, the excitation of C_{60} with 10 ns pulses leads to fragmentation by delayed ionization and C_2 emission, as well as other fragments—even for small intensities of about 2×10^6 W/cm^2 [15]. The use of electron spectroscopy in addition to the ion measurements allowed new questions to be posed, such as the impact of multi-electron dynamics and whether the ionization and the fragmentation dynamics be adequately modeled in the SAE picture [12]. This latter work resulted in other recent experimental and theoretical investigations, which concluded that both SAE and many-electron effects are important [8].

Trimetallic nitride template (TNT) endohedral metallofullerenes (EMFs), which consist of a trimetallic nitride moiety and a fullerene host, have also sparked broad interest in many fields—including materials chemistry, organic chemistry, biomedicine, biomedical chemistry, and molecular device design [20–23]. In addition to the fundamental photodynamics interest, EMFs carry the expectation or hope to act as radiotherapy agents to treat tumors while significantly reducing the X-ray dose for patients. Functional groups can be attached to the endohedral fullerene shell to bind the molecules to a specific site in order to deliver toxic, high-Z metal atoms, which are enclosed inside [24, 25]. Endohedral fullerenes have high stability, which is an inherent advantage for resisting biologically induced cage-opening [26].

http://dx.doi.org/10.5772/intechopen.70769

This is thought to help protect healthy tissues from the toxic metals. Monochromatic X-rays could be used to resonantly excite the high-Z atom, restricting the X-ray radiation damage to the cells that need to be treated, while leaving surrounding tissues largely unaffected by the radiation dose. In our case, we are interested in exploring the X-ray absorption and response of EMF with ultrashort X-rays.

2. FELs as tools for fullerene dynamics

A new class of intense and short-wavelength lasers, the FELs [27–32], has opened up new research opportunities for many scientific fields, from physics to chemistry, as well as matter under extreme conditions and biology. These vuv/X-ray lasers are accelerator-based tools, which are hybrid, as far as their attributes are concerned, between synchrotron facilities and typical tabletop lasers. FEL typically employs linear accelerators to drive relativistic electron beams through long undulators, characterized by alternating magnetic field, in order to produce intense radiation [32, 33]. **Figure 1** shows a schematic of the LCLS FEL undulator [34]. This type of insertion device, which was key for high-brightness third-generation synchrotron light sources, enable, if they are long enough, the production of ultra-intense vuv/X-ray radiation with femtosecond (fs) pulse duration [27, 30–32].

The use of FELs is unique despite the availability of fs and attosecond tabletop lasers to investigate ultrafast fullerenes and cluster dynamics. FELs add new attributes to tabletop lasers because they provide very high fluence (>10^{18} photon/pulse) in a wide, tunable photon energy range (10 eV–12 keV) that is not yet achievable with any tabletop laser. One of the essential attributes of the use of short wavelengths is to enable element selectivity, which permits selecting a specific atom in any system. Furthermore, short wavelengths allow for site selectivity, which

Figure 1. Schematic of the 33 m long LCLS FEL undulator [34].

thus targets a specific atomic shell in any system. This provides atomic-scale spatial resolution, which is very important for delineating the effects of each atom in a given system. Because FELs target atomic orbitals instead of molecular orbitals, they allow simple measurements of the response of inner-shell electrons (localized with each atom composing the molecule), compared to the complex response of the molecular orbitals composed of all valence electrons. Probing inner-shell electrons with short wavelengths allows efficient probing of physical and chemical phenomena in fullerenes or clusters from within, in an inside-out ionization.

Tabletop laser technology is also progressing to shorten the wavelength and increase the pulse energy, in order to enable study of photon-matter interactions as well—by ionizing or scattering from inner-shell electrons. Short-wavelength FELs also provide femtosecond pulse duration to study nuclear dynamics. Tabletop laser technology has already progressed to provide attosecond pulses, thus allowing the probing of electronic properties of given systems. In parallel, FEL scientists are also working hard at shortening their photon pulses; for example, the LCLS is aiming to produce ~600–400 attoseconds X-ray pulses in the fall of 2017. This new development will allow time-resolved studies of not only nuclear [5] but electronic molecular dynamics as well [35]. These new FEL-based xuv or X-ray photon sources initiate the photodynamics by core ionization. The initial charge is induced with high degrees of temporal and spatial localization, thus making the data analysis easier to handle. The new FEL family of ultrafast xuv/X-ray complement and extend the work carried out by tabletop lasers [36], xuv [37], synchrotron [38], or photon sources.

FELs are emerging photon tools that have been available since 2005, with the first vuv FLASH FEL at DESY in Germany [31] becoming available to scientists. The first X-ray FEL, the LCLS, was commissioned and available to scientists while being commissioned in 2009 [27, 30]. One of the most important attributes of FELs is their tunability within a wide frequency range. They currently span a photon energy range of about 10 eV to 12 keV, with pulse energies that exceed 3 mJ. This includes a repetition (rep) rate of up to 120 Hz, including a fs time scale where the pulse duration can be as short as 2–3fs and as long as 500 fs. The LCLS FEL aims to provide ~600–400 attoseconds in the soft X-ray regime by December 2017, which is an unprecedented progress. This new breakthrough will push the frontiers of science and may result in new science.

In addition to the X-ray LCLS FEL in the United States [27, 30] and the hard X-ray SACLA FEL in Japan's Riken laboratory [28], there are four vuv FEL counterparts: FLASH-1 and FLASH-2 at DESY [31], Germany; SCSS in Spring-8 Japan [28]; and FERMI in Trieste, Italy [29]. Other FELs that will be available this year, in 2017, are as follows: the large European XFEL project at DESY, Germany, will be available in the fall of 2017; a hard and soft X-ray FEL in South Korea called PAL-XFEL; and the SwissFEL at the Paul Scherrer Institute in Switzerland. Other FELS are either amidst planning in Sweden or in construction for the soft-X ray LCLS-II FEL in the United States.

Most FELs produce their photons via the self-amplified spontaneous emission (SASE) process, which is a stochastic process; thus, FEL laser pulses build up from noise and consist of a number of randomly spaced spikes of 1–5 fs duration within an energy envelope of about 15 eV—if it is not monochromatized—giving rise to *not* well-defined laser pulse profiles. The only exception so far is the FERMI FEL, which is a laser-seeded FEL that has similar pulse profiles compared to tabletop lasers. In addition, the FERMI FEL [29] produces a high-pulse energy,

http://dx.doi.org/10.5772/intechopen.70769

20–100 μJ, between 14 and 62 eV, with ~100 fs pulse duration, and the spectral width is typically $\Delta\lambda/\lambda \approx 1 \times 10^{-3}$ (FWHM). The combination of these parameters surpasses tabletop lasers. Progress toward the X-ray FEL pulse shape is happening; however, the LCLS has introduced a so-called self-seeding, where a tunable element—typically a crystal or grating monochromator—is inserted halfway through the undulators to filter a specific wavelength for further amplification. This provides higher stability and reduces the laser bandwidth from 20 to 0.5 eV (e.g., 0.75 eV bandwidth at 8450 eV photon energy). The XFEL, in Hamburg, Germany, which will be operating in the fall of 2017, will utilize a new technology based on superconducting linacs, in order to accelerate the electrons that produce the FEL photon pulses, with higher brightness and a higher repetition (rep) rate. This high rep rate of ~1 MHz enables highly differential measurements, such as e-ion-ion coincidence techniques that can delineate dynamics. These coincidence techniques will allow the examination of both electronic and nuclear dynamics, in real time, subsequent to photo-absorption, which is important for ultrafast chemistry.

If one were to compare tabletop lasers and FELs, he or she would state that they are complementary light sources, with their own advantages and challenges. The drawback with the tabletop lasers is the need to overcome their low-pulse intensity in the xuv-X-ray regime, as well as their short-range frequency tunability. They can provide weak intensity attosecond pulses, thus enabling electronic dynamics research in some cases [36]. Tabletop lasers deliver higher rep rates (~100 kHz) compared to FELs (120 Hz). However, as stated above, the XFEL in Germany will offer MHz rep rates in 2017. FELs, like synchrotron light sources, interact with inner-shell electrons instead of molecular orbitals. In fact, all FELs produce high-pulse intensity in the X-ray regime, which is ten orders of magnitude higher compared to synchrotron radiation. The broad tunability, with the use of monochromators, enables the exploration of resonance structures and enables wavelength-dependent experiments.

All short-wavelength light sources enable inner-shell ionization, which is followed by the Auger decay—thus, the absorption of one photon leads to the emission of two electrons, as shown in **Figure 2**. With ultra-intense FELs, the fluence is so high that multiple absorptions of photons occur, leading to sequential multiphoton ionization.

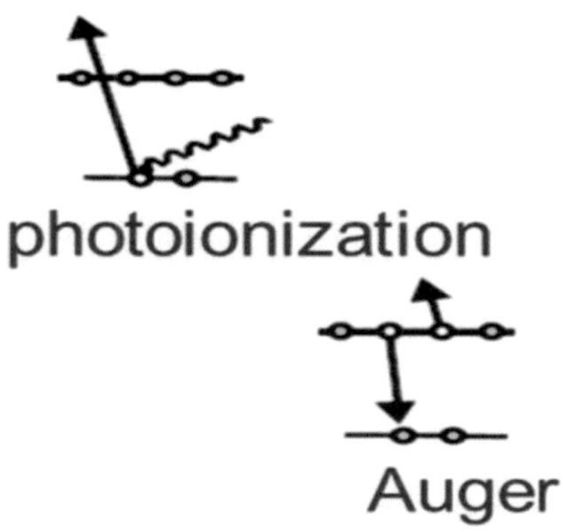

Figure 2. Schematic of the photoionization and Auger decay mechanism.

3. Ionization and fragmentation dynamics of C_{60} with the LCLS and FEL

The ionization and fragmentation of C_{60} were investigated experimentally and theoretically by using intense, fs, X-ray FEL pulses. The goal was to understand the response of a relatively large, strongly bonded molecule and to quantitatively understand the fs dynamical effects initiated by intense X-ray exposure. This is important to fundamental nonlinear physics research progress in general. It also directly impacts both the study of matter under extreme conditions and the area of biomolecular imaging. Imaging viruses and proteins, at an atomic spatial scale and on the time scale of atomic motion, requires quantitative understanding—which is best accomplished through spectroscopic measurements of a system that is simpler than a biomolecule—instead of scattering experiments. In other words, studying fullerenes, this is the first step toward understanding in a large system how a multiply ionized fullerene leads to slow electrons, which in turn initiate radiation damage in biological systems [39]. Therefore, the photoionization of C_{60} with intense FEL pulse is relevant to imaging of biomolecules, because such a study revealed the influence of processes that were not known or reported prior to our work [40].

The structure determination of biomolecules is done by imaging these systems using X-ray scattering experiments. These studies need photon brightness to obtain the images, which unfortunately induces electronic and structural radiation damage—thus altering the sample despite the use of short pulse durations [41]. Calculations have shown that high-intensity X-ray pulses trigger a cascade of damage processes in ferredoxin crystals, which are a metalloprotein of particular interest to the biology community [41, 42]. Furthermore, intense X-ray FEL pulses were found to modify the electronic properties of C_{60} on a fs time scale, based on observations of the diffraction of intense 32-fs X-ray pulses by a powder sample of crystalline C_{60} [41]. These radiation damages need to be evaluated, and spectroscopic measurements can reveal information that is not available with scattering experiments.

C_{60} was chosen as a benchmark molecule because of its chemically bonded carbon atoms, whose bond lengths and damage processes can emulate biomolecules. The fs experimental and theoretical C_{60} investigation dynamics was studied in the gas phase, with intense 485 eV photon energy to take advantage of the large photo-absorption cross section of carbon 1s electrons. We aimed to reach conditions in which approximately each C atom within a C_{60} molecule in the focus of X-ray pulse absorbs multiple photons during the X-ray pulse duration.

The experiment was carried out at the AMO hutch at the LCLS. The fullerenes (C_{60} and $Ho_3N@C_{80}$) were introduced into the vacuum through a heated oven source. For the C_{60} experiment, a 2m-long-magnetic-bottle time-of-flight spectrometer [43] was used to measure the ions and electrons that resulted from the ionization [40]. The spectrometer is shown in **Figure 3**, while a zoom of the interaction region is displayed in **Figure 4**; the oven appears as a cylinder from the top, the spectrometer's permanent magnet sticks to the left of the figure, and the spectrometer lens is shown to the right side of **Figure 4**.

We chose three pulse durations 4, 30, and 90 fs to explore the effect of the X-ray pulse duration on the C_{60} ionization. As explained with **Figure 2**, the inner-shell ionization will lead to

http://dx.doi.org/10.5772/intechopen.70769

Figure 3. Picture of the AMO hutch installation. The 2m-long-magnetic-bottle spectrometer, wrapped in aluminum foil, is shown sticking out from the interaction chamber.

Figure 4. Close-up view of the interaction region showing the nozzle part of the oven, the permanent magnet, and the spectrometer lens (see details in the text).

the Auger process, but with a high-fluence X-ray FEL pulse, this results in many photo- and Auger electrons due to the cyclic, multiphoton ionization until the pulse duration is over. Additionally, this photoionization process leads to secondary ionization of C_{60}, as well as of its fragment ions by the photo- and Auger electrons. The C_{60} molecule charges up to C_{60}^{8+} based on our data. The X-ray pulse is not homogeneous, as described in Section 2. Therefore, depending upon where C_{60} lands in the X-ray pulse, it can fragment in different scenarios. If C_{60} lands outside of the focus of the X-ray FEL pulse, it will fragment, as shown in **Figure 5**, into molecular ion strands with either a ring or linear structure. **Figure 5** shows the mass-to-charge ratio (M/Q), displaying the parent C_{60} ionized up to q = 8+ along with the molecular ions C_m^{+}, m = 1–11.

If C_{60} lands in the focus of the FEL pulse, because of the short C—C bond lengths, it fragments via the Coulomb repulsion into small molecular carbon ions and atomic carbon charge-state distribution from C^{+} to C^{6+}. Therefore, C can be fully stripped with 90 fs pulse duration [40] as shown in **Figure 6**. In this case, the pulse is intense and long enough for the C atom to undergo at least three cyclic photoionization-Auger decays, leading to C^{6+}.

As shown in **Figure 6**, the charge-state distribution depends upon the pulse duration. At the shorter pulse duration we used (4fs), the highest charge state obtained is C^{5+} indicating that at least two cyclic photoionization-Auger decays occurred [40, 44]. Our measurements (light color data) were compared with calculation (dark color data) in order to understand the different physical and chemical effects, giving rise to the observed charge-state distribution.

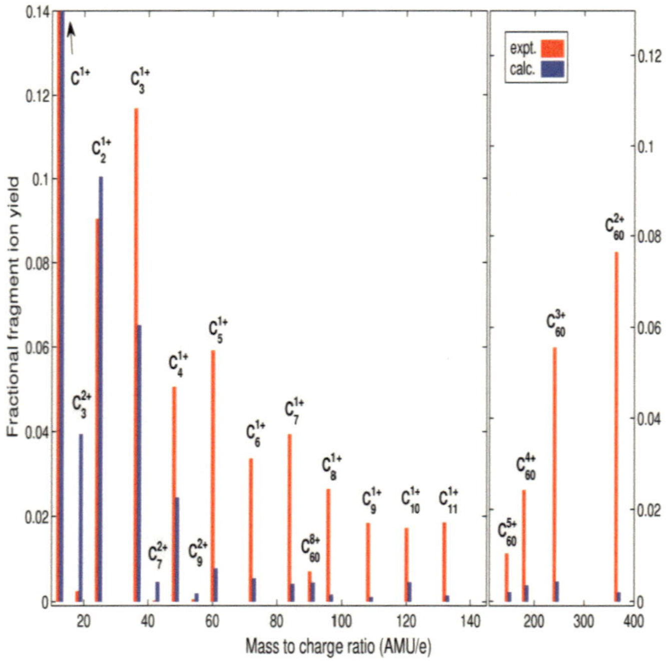

Figure 5. Time-of-flight spectrum showing the fragmentation of C_{60} with mid-fluence LCLS X-ray pulses. Light color data are experiment while dark color data are theory.

http://dx.doi.org/10.5772/intechopen.70769

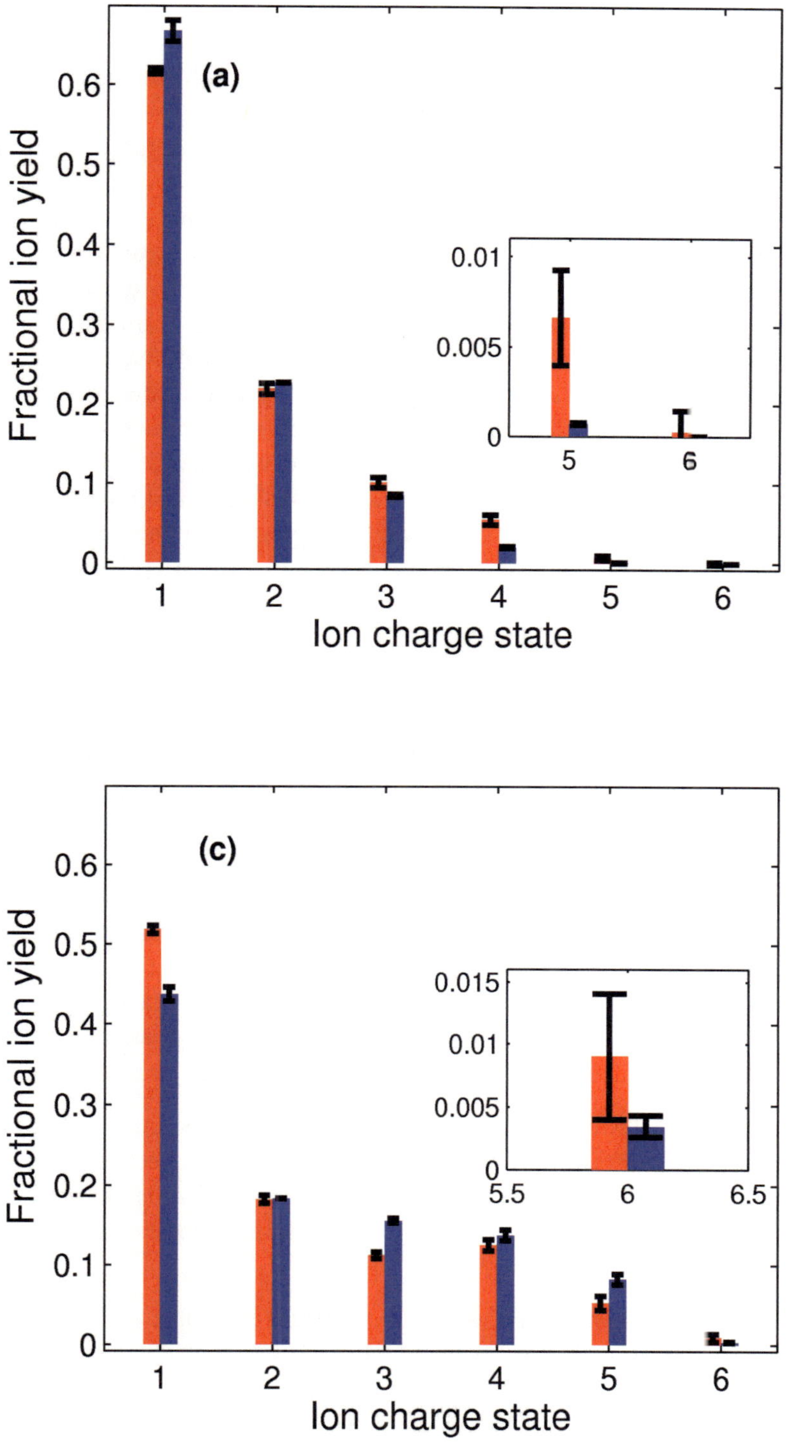

Figure 6. Atomic carbon charge-state distribution resulting from the Coulomb explosion of C_{60} under high-fluence X-ray FEL pulses. (a) corresponds to data taken with 4f pulse duration while (c) corresponds to data taken with 90 fs pulse duration.

3.1. Discussion

With current computer technology, it is not possible to carry out ab initio calculations via quantum mechanical methods. The best that can be done is to develop classical physics models that describe the dynamics of atoms/ions and electrons that appear in the continuum after photoionization. These models include, for example, the rate equations of the ionization and the C ionization cross sections that are directly resulting from quantum mechanical calculations [45, 46]. The model used in this FEL-based C_{60} work consists of a mixed molecular dynamics (MD) Monte Carlo tool based on treating C atoms [47]. The real-space dynamics of atoms, ions, and the (quasi-) free electrons resulting from photoionization and Auger decay is described by Newtonian mechanics. In this methodology, C_{60} is modeled as 60 individual carbon atoms. The atoms' electronic configuration is tracked and changed stochastically during each time step (0.8 attosecond) by using a Monte Carlo scheme. During the many ionization events, new electrons that are released from their bound atomic orbitals are treated classically, with the appropriate energies, as they "appear" in the system. The C atoms were held together by a fullerene-specific classical Brenner force field, and the charges interacted via Coulomb forces. In order to mimic a molecule, the model needed to contain several physical and chemical processes that were revealed through the experiment. **Figure 6** shows a good agreement between the model and the experiment, but in the initial comparison, they did not agree. The model originally predicted more abundant C ion charge states, which revealed that there was *strong recombination* between the released electrons from photoionization and the ions after the FEL pulse ends.

The comparison between the model and the experimental data not only led to the interpretation of our experiment but also led to solid improvement in the MD calculation. Specifically, the model had to include several physical and chemical processes: (1) bond-breaking in C_{60} was modeled by setting the bonding force field to zero between two C-charged ions; (2) the secondary ionizations arising from free electrons colliding with and ionizing atoms/ions was included; (3) the molecular Auger effect consisted of sharing the energy and using two carbon atoms in the release of the two electrons. This was taken into account by removing the ejected electron from the L-shell of a neighboring C atom, while the 1s vacancy of an ion was filled up with its own L-shell electron; and (4) the recombination of a classically trapped electron if it were no longer delocalized among several C ions but instead localized to only one C ion.

Figure 7 shows the effect of the MD modeling, where we compare the data and the model for two pulse durations, 30 and 90 fs. As can be seen, there is no effect with the two different pulse durations, which thus revealed that the dynamics might already be over with the use of a 30 fs X-ray pulse. This means that the physical and chemical processes included might have occurred within 30 fs.

In order to explain the comparison between the model and the data, we display in **Figure 7** the C ion yield *difference* between the model and the experiment on the Y-axis. The smallest value of the bars corresponds to the best agreement between the model and the experiment. Each yield corresponds to the sum of the charge-state distribution (C^{1+}–C^{6+}) shown in **Figure 7**. The X-axis displays bars that correspond to the several different physical and chemical processes, which were included in the model. As described above, each of these processes were included

http://dx.doi.org/10.5772/intechopen.70769

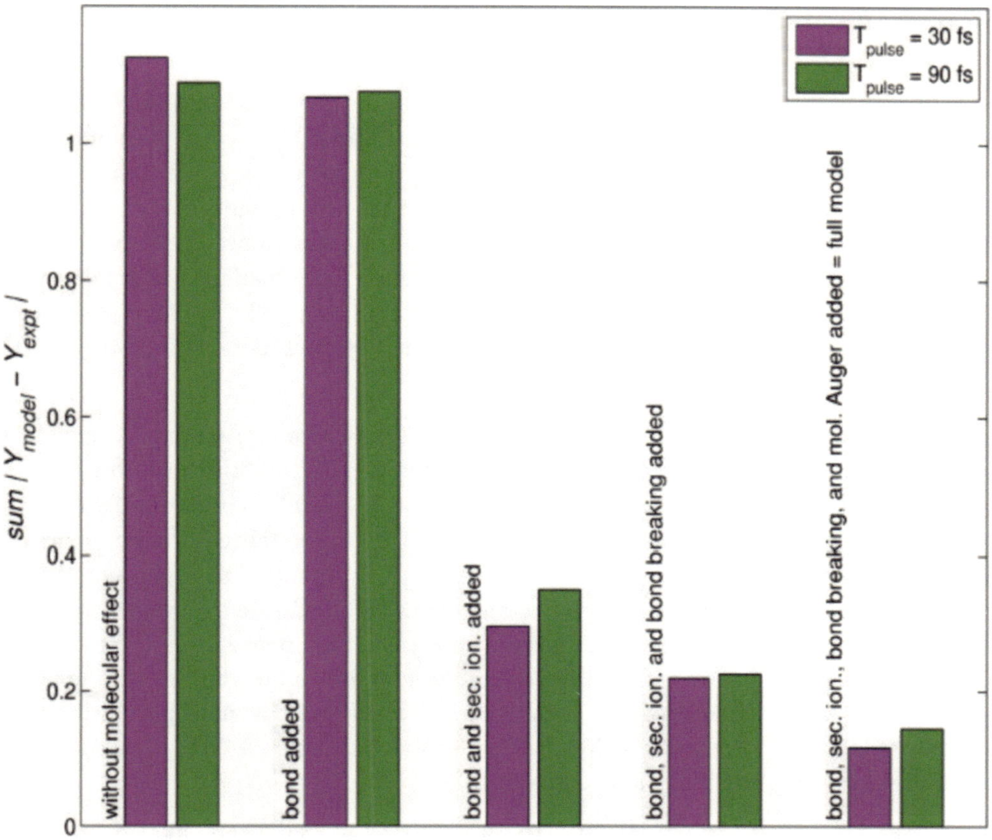

Figure 7. Comparison between the experimental data and the model (see text for details).

one at a time to display their importance. As can be seen, molecular effects and molecular bonds slightly contributed to a better agreement. The dramatic effect is obtained in the case of the secondary ionization of C_{60} by the photo- and Auger-electrons. This effect is weak in small, isolated molecules and van der Waals clusters; it is completely absent in atoms. As can be seen, the addition of molecular bond-breaking and molecular Auger improved even more the model.

This combined theoretical and experimental work illustrates the successful use of *classical mechanics* to describe all moving particles in C_{60}. The work clearly revealed the influence of processes not previously suspected or reported. The impact we aimed to achieve was also realized because this fullerene spectroscopic work quantitatively demonstrated electronic damage due to photoelectron and Auger electrons interacting with the ions as a secondary ionization effect. Our results were corroborated with recent separate calculations [39]. Finally, one of the goals of the MD model was to build a new approach that would scale with larger systems, such as biomolecules. This work demonstrated that the modeling [40, 44] coached by experiment was successful and can be applicable for X-ray interactions with any extended system—even at higher X-ray dose rates that are expected with future FEL sources, such as the soon-to-be available XFEL in Germany and with the future LCLS-II.

4. Ionization and fragmentation dynamics of $Ho_3N@C_{80}$

The C_{60} work described above led us to consider exploring increased complexity by choosing the interaction of endohedral fullerenes (also called doped fullerenes) with FELs. These systems are even more intriguing than C_{60}, because they host a moiety that ranges from an atom to a molecule. Nothing was known about their structure or dynamics when excited with X-ray FEL. As mentioned in the introduction, these nanoscale systems have received attention in part because they can be used for applications ranging from medical usage [48] to drug delivery, as well as their possible use for quantum computing [49]. Our interest, however, stems from exploring a fundamental point of view, the fragmentation of $Ho_3N@C_{80}$ induced by short, X-ray pulses from the LCLS FEL.

The experiment was carried out at the AMO hutch using a time-of-flight spectrometer [50] for detecting the ions produced in the interaction of the endohedral fullerenes with the LCLS pulses. The experiment on $Ho_3N@C_{80}$ was carried out with 1530 eV in order to selectively target the ionization from a specific shell, which was the Ho 3d. The pulse duration of the X-ray pulse was 80 fs with a pulse energy of about 6.7×10^{15} photons/cm^2. This experiment had less fluence than the experiment on C_{60} by two orders of magnitude due to a different transport of the photon beam through the optics [50]. We estimate that the fluence used in the $Ho_3N@C_{80}$ is only ¼ of the fluence used in C_{60} [40], thus resulting in only a few multiphoton ionization cycles. **Figure 8** shows three time-of-flight spectra that result from the multiphoton ionization processes of $Ho_3N@C_{80}$ with intense X-ray FEL. We show in **Figure 8** three panels that focus on the singly, doubly, and triply ionized parent.

The top panel, which spans the mass-to-charge ratio (M/Q) from 350 to 1300, depicts structures attributed to multiply charged fullerenes' parent ions from singly charged $Ho_3N@C_{80}^+$ to quintuply charged $Ho_3N@C_{80}^{5+}$ (albeit weak). This clearly shows that the fluence for this experiment is weaker than for the C_{60} experiment we described above since in **Figure 5** the parent ions were charged up to C_{60}^{8+}. It is also clear that the atomic Ho^+ ion has the highest yield compared to all other ion fragments. This must arise from the fragmentation of the encapsulated Ho_3N moiety—thus, indicating that the three multiply charged Ho atoms, which were selected to be the most ionized compared to N or C atoms, freed themselves from the C_{80} cage. The middle panel focuses on the doubly and triply ionized parent molecule along with doubly ionized fullerene molecules that lost C dimers such as $Ho_3N@C_{78}^{2+}$, $Ho_3N@C_{70}^{2+}$, and $Ho_3N@C_{50}^{2+}$. The loss of C dimers was also observed previously with tabletop experiments [2]; however, the shrunk cage size to C_{50} was never previously observed. The bottom panel shows the triply and quadruply charged parent fullerenes, along with triply ionized parent fullerenes that lost C dimers, namely, $Ho_3N@C_{78}^{3+}$, $Ho_3N@C_{76}^{3+}$, $Ho_3N@C_{70}^{3+}$, and $Ho_3N@C_{50}^{3+}$.

Figure 9 shows other fragments that occur in the 160–240 M/Q range. The prominent signal is the atomic Ho^+ ion which escaped the cage. Other fragment ions are C molecular ion chains or rings such as C_{15}^+ and C_{17}^+ along with surprising new fragments such as HoC_2^+, $HoCN^+$, HoC_4^+, and HoC_3N^+ [51].

http://dx.doi.org/10.5772/intechopen.70769

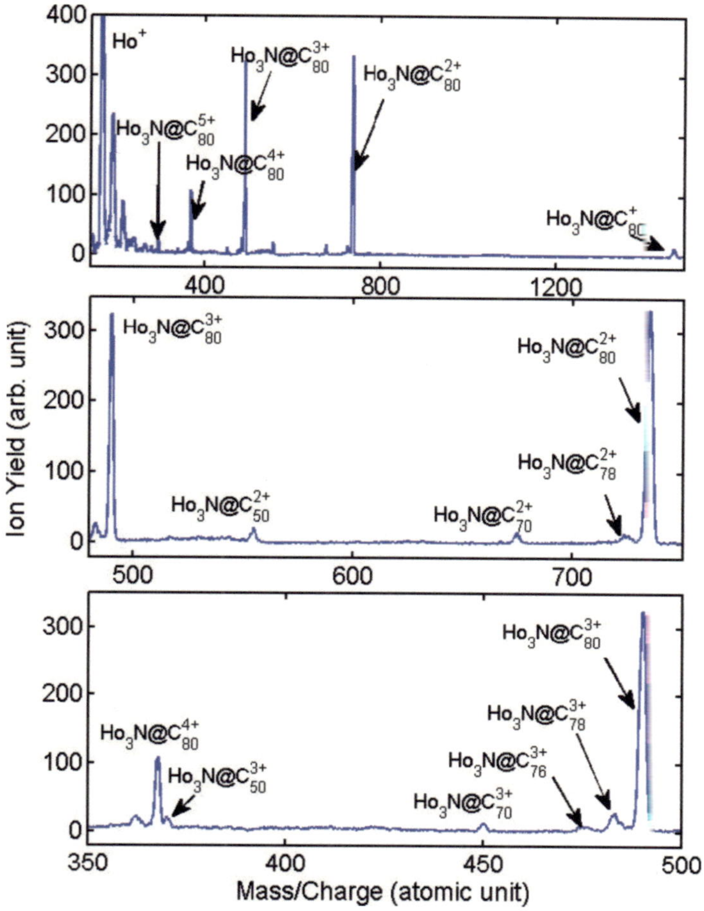

Figure 8. Ion yield M/Q spectra shown in three panels. The top panel depicts a wide M/Q fragment ions, while the middle and bottom fragments focus on doubly and triply charged parent ions (see text for details).

4.1. Discussion

This experiment was carried out in the low-fluence regime, since we estimated that about eight photons were absorbed by $Ho_3N@C_{80}$. In the C_{60} experiment, described in Section 3, we appraised that about 180 photons were absorbed per C_{60} molecule. The photoionization with 1530 eV leads to the absorption cross section of Ho to be about 1.55 Mb, while that of C was about 0.013 Mb. Our interpretation of the interaction of $Ho_3N@C_{80}$ with 1530 eV photon energy was that the Ho atom charges up and gets multi-ionized due to the cyclic photoionization and Auger decay [51]. In doing so, it grabs electrons from the carbon cage via electron transfer between the Ho and the cage—resulting in the whole parent molecule charging up and reaching at least $Ho_3N@C_{80}^{5+}$, as observed in **Figure 8**. We assumed that as the carbon cage charges up, it would become unstable and will break apart, thus leading to molecular

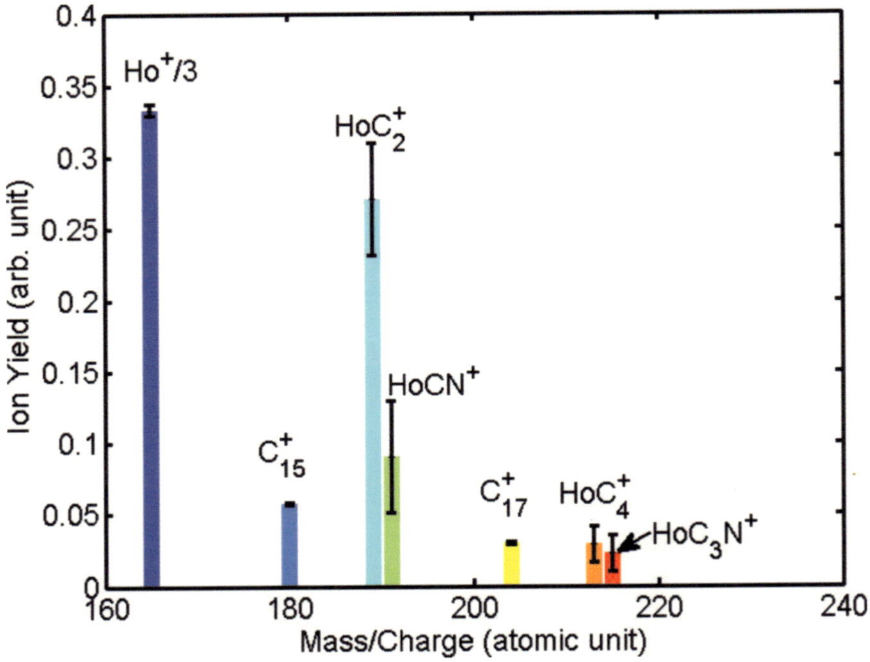

Figure 9. Ion spectrum displaying M/Q focusing on Ho ion and the Ho-based molecular fragments ion (see text for details).

fragment ions. The Ho atoms are about ten times heavier than the C atoms, and, therefore, we assume that they will not move faster than the carbon cage. We clearly have evidence of bond-breaking and bond-forming since we observed in **Figure 9** the following fragments: HoC_2^+, $HoCN^+$, HoC_4^+, and HoC_3N^+. It is unclear if the moiety first breaks into three Ho atoms and the N atoms and then the carbon cage fragments or if the reverse occurs. It is also unclear how the new bonds have formed. This is to be determined by future time-resolved experiments that might track the ionization and fragmentation dynamics and decipher the mechanisms leading to the final ionic states we observed. Additionally, we hope that our work will stimulate the development of molecular dynamics simulations suitable for endohedral fullerenes and for even larger molecules exposed to intense XFEL.

5. Measuring time-resolved dynamics using pump-probe techniques with FELs

Pump-probe spectroscopy techniques, championed by tabletop lasers [36], allow the measurement of dynamics for any system—from atoms to fullerenes and from solids to biological specimens. They are being used extensively, and the ultimate goal is to determine the motions and locations of nuclei and electrons and to determine the energy flow and charge transfer in systems. Recording these motions and making "molecular movies" using pump-probe spectroscopy seem reachable in the near future due to the new technologies.

Pump-probe techniques with FELs offer similar opportunities to measure physical and chemical changes in molecules at an atomic spatial resolution (on the time scale of atomic motion). With attosecond, the goal is to also measure the electronic motion. These techniques can be used for the study of fullerenes, in order to tackle fundamental questions such as, *how do the atoms in the fullerenes move after the photon energy is deposited in the fullerenes?* How do the bonds between the atoms that make the fullerenes break, or what are the pathways for the induced atomic motion in fullerenes? These questions can be asked and answered with FELs using pump-probe techniques. As in standard pump-probe work, to capture the dynamics, the pump pulse initiates the motions, and a probe pulse detects the changes using as many time delays as needed between the pump and probe pulses, ideally in a wide time scale. In order to achieve this goal, one needs two pulses, which can be generated in a few ways: (1) the accelerator scientists have developed several methods, but the most recent one cuts fresh slices from the electron bunch. They manipulate the electron bunch before it enters a split undulator, so that only its tail is lasing, in order to produce the first X-ray pulse. Next, they delay the electron bunch in order to acquire a time delay and spoil the electron bunch orbit further, by having the head of the bunch lasing and thus using a fresh slice from the electron bunch to produce the second photon pulse [52]. This scheme is also capable of providing two X-ray colors [52]; (2) the FEL pulse can be split into two X-ray pulses with an X-ray split and delay tool [53]. These two schemes were used successfully in experiments in the mode of X-ray pump-X-ray probe; and (3) a third scheme utilizes an X-ray pulse from the FEL as either the pump or the probe, and it is paired with a short tabletop pulse laser [54] (IR or UV). All of these schemes have been used in various FELs.

The FEL-based experiments, paired most of the time with pump-probe techniques, are carried out by using various types of spectrometers or imaging detectors for absorption experiments or for diffractive scattering experiments, respectively. The latter holds the promise to achieve imaging of molecules in the gas phase. Double VMI spectrometers paired with X-ray/IR, for a pump-probe scheme, are also used to measure the electrons—as well as the ions using the ion-ion coincidence techniques. These differential techniques examine both electronic and nuclear dynamics following the interaction of fullerenes with FEL X-ray pulses and also resolve the transient electronic rearrangement that accompanies photoionization [55]. The current repetition (rep) rate for the operational FEL facilities is at most 120 Hz (lower than the current tabletop repetition rate of 1–100 KHz), which does not allow easy FEL-based electron-ion-ion coincidence studies—although some work has been done in this highly differential mode [56]. However, the new XFEL in Hamburg, Germany, which will be completed by the end of this year, will offer ~1 MHz rep rate—thus allowing the measurement of e-ion-ion coincidences to track and precisely determine the decay channels and time it takes for the ultrafast nuclear dynamics to occur. It seems that the dream of making molecular movies is reachable in the next few years.

6. Conclusions

The investigation of the ionization and fragmentation of fullerenes with FELs is at its infancy. This work reported on the first two spectroscopic experiments on C_{60} and $Ho_3N@C_{80}$ using ion spectroscopy but much more needs to be accomplished. These first experiments need to be followed by time-resolved studies to delineate the nuclear dynamics, including electron transfer, using e-ion-ion coincidence techniques or using diffraction techniques and imaging. In addition, the next generation of attosecond FELs will allow the exploration and hopefully understanding of electron migration between atoms using photoelectron spectroscopy. Finally, the work done on C_{60} was conclusive because of the close interaction with theories and modeling. We hope that the reported work will stimulate theorists to tackle the many-body interactions resulting from the interaction of fullerenes with femtosecond or attosecond X-ray or vuv FELs.

Acknowledgements

We thank all of the authors of references [40, 44, 51]. We also thank Ms. Rachel Tshonas for her help with the manuscript. This work was funded by the Department of Energy, Office of Science, Basic Energy Sciences (BES), Division of Chemical Sciences, Geosciences, and Biosciences under grant no. DE-SC0012376.

Author details

Nora Berrah

Address all correspondence to: nora.berrah@uconn.edu

Physics Department, University of Connecticut, Storrs, CT, USA

References

[1] Supramolecular Chemistry of Fullerenes and Carbon Nanotubes. In: Martin N, Nierengarten JF, editors. Wiley-VCH Verlag GmbH & Co. KGaA; 2012. DOI: 10.1002/9783527650125

[2] Johansson JO, Campbell EEB. Probing excited electronic states and ionisation mechanisms of fullerenes. Chemical Society Reviews. 2013;**42**:5661-5671. DOI: 10.1039/C3CS60047E and references therein

[3] Feng M, Zha J, Petek H. Atomlike, hollow-core-bound molecular orbitals of C_{60}. Science. 2008;**320**:359-362 and references therein. DOI: 10.1126/science.1155866

[4] Mignolet B, Johansson JO, Campbell EEB. Probing rapidly ionizing super-atom molecular orbitals in C_{60}: A computational and femtosecond photoelectron spectroscopy study. Chemphyschem. 2013;**14**:3332-3340. DOI: 10.1002/cphc.201300585 and references therein

[5] Krausz F, Stockman MI. Attosecond metrology: from electron capture to future signal processing. Nature Photonics. 2014;**8**:205-213. DOI: 10.1038/nphoton and references therein

[6] Zhao SF, Le AT, Jin C, Wang X, Lin CD. Analytical model for calibrating laser intensity in strong-field-ionization experiments. Physical Review A. 2016;**93**:023413. DOI: 10.1103/PhysRevA.93.023413

[7] Bhardwaj VR, Corkum PB, Rayner DM. Internal laser-induced dipole force at work in C_{60} molecule. Physical Review Letters. 2003;**91**:203004 and references therein. DOI: 10.1103/PhysRevLett.91.203004

[8] Huismans Y, et al. Macro-atom versus many-electron effects in ultrafast ionization of C_{60}. Physical Review A. 2013;**88**:013201. DOI: 10.1103/PhysRevA.88.013201

[9] Campbell EE, et al. From above threshold ionization to statistical electron emission: The laser pulse-duration dependence of C_{60} photoelectron spectra. Physical Review Letters. 2000;**84**:2128

[10] Kjellberg M, et al. Momentum-map-imaging photoelectron spectroscopy of fullerenes with femtosecond laser pulses. Physical Review A. 2010;**81**:023202. DOI: 10.1103/PhysRevA.81.061602

[11] Bohl E, et al. Relative photoionization cross sections of super-atom molecular orbitals (SAMOs) in C_{60}. The Journal of Physical Chemistry A. 2015;**119**:11504-11508. DOI: 10.1021/jp408147f

[12] Shchatsinin I, et al. C_{60} in intense short pulse laser fields down to 9 fs: excitation on time scales below e-e and e-phonon coupling. The Journal of Chemical Physics. 2006; **125**:194320. DOI: http://dx.doi.org/10.1063/1.2362817

[13] Campbell EEB, Hoffmann K, Rottke H, Hertel IV. Sequential ionization of C_{60} with femtosecond laser pulses. The Journal of Chemical Physics. 2001;**114**:1716. DOI: http://dx.doi.org/10.1063/1.1584671

[14] Tchaplyguine MHK et al. Ionization and fragmentation of C_{60} with sub-50 fs laser pulses. The Journal of Chemical Physics. 2000;**112**:2781. DOI: http://dx.doi.org/10.1063/1.480852

[15] Lebeault MA, et al. Decay of C_{60} by delayed ionization and C_2 emission: Experiment and statistical modeling of kinetic energy release. The Journal of Chemical Physics. 2012;**137**:054312. DOI: http://dx.doi.org/10.1063/1.4737926

[16] Li H, et al. Coherent electronic wave packet motion in C_{60} controlled by the waveform and polarization of few-cycle laser fields. Physical Review Letters. 2015;**114**:123004. DOI: 10.1103/PhysRevLett.114.123004

[17] Xiong H, Mignolet B, Fang L, Osipov T, Thomas T, Wolf JA, Sistrunk E, Gühr M, Remacle R, Berrah N. The role of super-atom molecular orbitals in doped fullerenes in a femtosecond intense laser field. Scientific Reports. 2017;**7**:121

[18] Li H, et al. Transition from SAMO to Rydberg state ionization in C_{60} in femtosecond laser fields. The Journal of Physical Chemistry Letters. 2016;**7**:4677-4682. DOI: 10.1021/acs.jpca.5b11713

[19] Jaroń-Becker A, Becker A, Faisal FHM. Single-active-electron ionization of C_{60} in intense laser pulses to high charge states. The Journal of Chemical Physics. 2007;**126**:124310. DOI: 10.1021/acs.jpca.5b11713

[20] Fatouros PP et al. In vitro and in vivo imaging studies of a new endohedral metallofullerene nanoparticle. Radiology. 2006;**240**:756. DOI: 10.1148/radiol. 2403051341

[21] Fillmore HL, et al. Conjugation of functionalized gadolinium metallofullerenes with IL-13 peptides for targeting and imaging glial tumors. Nanomedicine. 2011;**6**:449-458. DOI: 10.2217/nnm.10.134

[22] Popov AA, Yang S, Dunsch L. Endohedral fullerenes. Chemical Reviews. 2013;**113**:5989-6113. DOI: 10.1021/cr300297r

[23] Zhang J, Stevenson S, Dorn HC. Trimetallic nitride template endohedral metallofullerenes: Discovery, structural characterization, reactivity, and applications. Accounts of Chemical Research. 2013;**46**:1548. DOI: 10.1021/ar300301v

[24] Cagle DW, Kennel SJ, Mirzadeh S, Alford JM, Wilson LJ. In vivo studies of fullerene-based materials using endohedral metallofullerene radiotracers. Proceedings of the National Academy of Sciences of the United States of America. 1999;**96**:5182-5187

[25] Meng J, Liang X, Chen X, Zhao Y. Biological characterizations of $[Gd@C_{82}(OH)_{22}]_n$ nanoparticles as fullerene derivatives for cancer therapy. Integrative Biology. 2013;**5**:43. DOI: 10.1039/C2IB20145C

[26] Stevenson S et al. Small-bandgap endohedral metallofullerenes in high yield and purity. Nature. 1999;**401**:55. DOI: 10.1038/43415

[27] Emma P, et al. First lasing and operation of an ångstrom-wavelength free-electron laser. Nature Photonics. 2010;**4**:641-647. DOI: 10.1038/nphoton.2010.176

[28] Ishikawa T, Allaria E, et al. Ultrafast X-ray pulse characterization at free-electron lasers. Nature Photonics. 2012;**6**:540-544. DOI: 10.1038/nphoton.2012.276

[29] Allaria E, et al. Two-stage seeded soft-X-ray free-electron laser. Nature Photonics. 2012;**6**:699-704. DOI: 10.1038/nphoton.2013.277

[30] Berrah N, Bucksbaum PH. The Ultimate X-ray Machine. Scientific American. 2014;**310**:64-71. DOI: 10.1038/scientificamerican0114-64

[31] Ackermann W et al. Operation of a free-electron laser from the extreme ultraviolet to the water window. Nature Photonics. 2007;**1**:336-342. DOI: 10.1038/nphoton.2007.76

[32] Bucksbaum P, Berrah N. Brighter and faster: The promise and challenge of the X-ray free-electron laser. Physics Today. 2015;**68**:25-32. DOI: 10.1063/PT.3.2845

[33] Berrah N, Bucksbaum PH. The ultimate X-ray machine. Scientific American Special Collector's Edition. 2015;**54**:7

[34] Figure 1 from SLAC site: https://portal.slac.stanford.edu/sites/lcls_public/Pages/Default.aspx

[35] Kraus PM, et al. Measurement and laser control of attosecond charge migration in ionized iodoacetylene. Science. 2015;**350**:790-795. DOI: 10.1126/science aab2160

[36] Leone SR, McCurdy CW, Burgdoerfer J, Cederbaum LS, Chang Z, Dudovich N, Feist J, Greene CH, Ivanov M, Kienberger R, Keller U, Kling MF, Loh ZH, Pfeifer T, Pfeiffer AN, Santra R, Schafer K, Stolow A, Thumm U, Vrakking MJJ. Charge migration in molecules. Nature Photonics. 2014;**8**:162-166. DOI: 10.1038/nphoton.2014.48

[37] Calegari F, et al. Ultrafast electron dynamics in phenylalanine initiated by attosecond pulses. Science. 2014;**346**:336-339. DOI: 10.1126/science.1254061

[38] Sann H, Havermeier T, et al. Imaging the temporal evolution of molecular orbitals during ultrafast dissociation. Physical Review Letters. 2016;**117**:243002. DOI: 10.1103/PhysRevLett.117.243002

[39] Stumpf V, Gokhberg K, Cederbaum LS. The role of metal ions in X-ray-induced photochemistry. Nature Chemistry. 2016;**8**:237-224. DOI: 10.1038/nchem.2429

[40] Murphy BF, et al. Femtosecond X-ray-induced explosion of C_{60} at extreme intensity. Nature Communications. 2014;**5**:4281. 1-9 and references therein. DOI: 10.1038/ncomms5281

[41] Abbey B, et al. X-ray laser–induced electron dynamics observed by femtosecond diffraction from nanocrystals of Buckminsterfullerene. Science Advances. 2016;**2**:1-5 and references therein. DOI: 10.1126/sciadv.1601186

[42] Nass K, et al. Indications of radiation damage in ferredoxin microcrystals using high-intensity X-FEL beams. Journal of Synchrotron Radiation 2015;**22**(1):1-14. ISSN: 1600-5775 and references therein. DOI: 10.1107/S1600577515002349

[43] Eland J, Vieuxmaire O, Kinugawa T, Lablanquie P, Hall R, Penent F. Complete two-electron spectra in double photoionization: The rare gases Ar, Kr, and Xe. Physical Review Letters. 2003;**90**:053003. DOI: 10.1103/PhysRevLett.90.053003

[44] Berrah N, Fang L, Osipov T, Jurek Z, Murphy BF, Santra R. Emerging photon technologies for probing ultrafast molecular dynamics. Faraday Discussions. 2014;**171**(1):471-485. DOI: 10.1039/C4FD00015C

[45] Son SK, Santra R. Investigating dynamics of complex system irradiated by intense X-ray free electron laser pulses. Journal of Phys. Conf. Series; 2011 XATOM—an integrated toolkit for X-ray and atomic physics (CFEL, DESY)

[46] Son SK, Young L, Santra R. Impact of hollow-atom formation on coherent X-ray scattering at high intensity. Physical Review A. 2011;**83**:033402-033413. DOI: 10.1103/PhysRevA.83.033402

[47] Jurek Z, Faigel G, Tegze M. Dynamics in a cluster under the influence of intense femtosecond hard X-ray pulses. European Physical Journal D: Atomic, Molecular, Optical and Plasma Physics. 2004;**29**:217-229. DOI: 10.1140/epjd/e2004-00033-3

[48] Melanko JB, Pearce M, Salem A. Nanotubes/nanorods/nanofibers/fullerenes for nanoscale drug delivery. Nanotechnology in Drug Delivery; Springer Science/American Association for Pharmaceutical Scientists (AAPS). 2007;105-127

[49] Harneit W, et al. Room Temperature electrical detection of spin coherence in C_{60}. Physical Review Letters. 2007;**98**:216601. DOI: 10.1103/PhysRevLett.100.199904

[50] Hoener M, Fang L, Kornilov O, Gessner O, Pratt S, Guehr M, Kanter E, Blaga C, Bostedt C, Bozek J, et al. Ultraintense X-ray induced ionization, dissociation, and frustrated absorption in molecular nitrogen. Physical Review Letters. 2010;**104**:253002. DOI: 10.1103/PhysRevLett.104.253002

[51] Berrah N, Murphy B, Xiong H, Fang L, Osipov T, Kukk E, Guehr M, Feifel R, Petrovic VS, Ferguson KR, Bozek JD, Bostedt C, Frasinski LJ, Bucksbaum PH, Castagna JC. Femtosecond X-ray induced fragmentation of fullerenes. Journal of Modern Optics. 2015;**3**(4):390-401. DOI: 10.1080/09500340.2015.1064175

[52] Lutman I, Maxwell T, MacArthur J, Guetg M, Berrah N, Coffee R, Ding Y, Huang MZA, Moeller S, Zemella J. Fresh-slice multicolour X-ray free-electron lasers. Nature Photonics. 2016;**10**:745. DOI: 10.1038/nphoton.2016.201

[53] Berrah N, Fang L, Murphy B, Kukk E, Osipov T, Coffee R, Ferguson K, Xiong H, Castagna JC, Petrovic V, Montero S, Bozek J. A two mirrors X-ray pulse split and delay instrument for femtosecond time resolved investigations at the LCLS free electron laser facility. Optics Express. 2016;**24**(11):11768-11781. DOI: 10.1364/OE.24.011768

[54] McFarland B, Farrell J, Miyabe S, Tarantelli F, Aguilar A, Berrah N, Bostedt C, Bozek J, Bucksbaum P, Castagna J, et al. Ultrafast X-ray Auger probing of photoexcited molecular dynamics. Nature Communications. 2014;**5**:4235. DOI: 10.1038/ncomms5235. DOI:10.1038/ncomms5235

[55] Erk B, et al. Imaging charge transfer in iodomethane upon X-ray photoabsorption. Science. 2014;**345**:288-291 and references therein. DOI: 10.1126/science.1253607

[56] Moshammer SK, Jaeschke E, et al. Synchrotron light sources and free-electron lasers. Springer International Publishing Switzerland. 2016;1291-1322. DOI: 10.1007/978-3-319-14394-1-26

Chapter 5

Structural and Vibrational Properties of C_{60} and C_{70} Fullerenes Encapsulating Carbon Nanotubes

Hassane Chadli, Fatima Fergani and
Abdelali Rahmani

Additional information is available at the end of the chapter

http://dx.doi.org/10.5772/intechopen.71246

Abstract

Carbon nanotubes (CNTs) can encapsulate small and large molecules, including C_{60} and C_{70} fullerenes (so-called carbon peapods). The challenge for nanotechnology is to achieve perfect control of nanoscale-related properties, which requires correlating the parameters of synthesis process with the resulting nanostructure. For that purpose, note every conventional characterization technique is suitable, but Raman spectroscopy has already proven to be. First, the different possible configurations of C_{60} and C_{70} molecules inside CNTs are reviewed. Therefore, the following changes of properties of the empty nanotubes, such as phonon modes, induced by the C_{60} and C_{70} filling inside nanotube are presented. We also briefly review the concept of Raman spectroscopy technique that provides information on phonon modes in carbon nanopeapods. The dependencies of the Raman spectrum as a function of nanotube diameter and chirality, fullerene molecules configuration and the filling level are identified and discussed. The experimental Raman spectra of fullerenes and fullerenes peapods are discussed in the light of theoretical calculation results. Finally, the variation of the average intensity ratio between C_{60} and C_{70} Raman-active modes and the nanotube ones, as a function of the concentration molecules, are analyzed, and a general good agreement is found between calculations and measurements.

Keywords: fullerene, C_{60}, C_{70}, carbon nanotubes, peapods, Raman spectroscopy

1. Introduction

Since Kroto *et al.* discovered C_{60} by laser vaporizing graphite into a helium stream in September 1985, fullerenes are at the heart of nanotechnology [1]. Other fullerenes were discovered shortly afterward with more and fewer carbon atoms. Since the discovery of multiwalled carbon nanotubes and single-walled carbon nanotube (SWCNT) in 1991 [2, 3], fullerenes and carbon nanotube systems have attracted significant attention from the scientific community. A remarkable property of SWCNT is its ability to have been filled with various fullerenes and metallofullerenes, fullerenes adducts, metal complexes, and other small molecules. These fillings are highly dependent on the nanotube diameter and the inserted molecule size, so that even small changes in SWCNT diameter can alter the geometry of fullerene arrays. This class of hybrid materials has been dubbed as "peapods" (C_{60}@SWCNT and C_{70}@SWCNT), reflecting structural similarities to real peapods. After the discovery of C_{60} peapods by Smith et al. in 1998 [4], many experimental studies clearly evidenced the existence of various fullerenes like C_{70}, C_{76}, and C_{80} inside SWCNTs [5–7]. However, these materials represent a new class of a hybrid system between fullerenes and SWCNTs where the encapsulated molecules peas and the SWCNT pod are bonded through van der Waals interactions. Using high-resolution transmission electron microscopy (HR-TEM) experiments, the peapods are clearly observed, as seen in **Figure 1**, and organized into bundles [6–8].

The physicochemical properties of the fullerene molecules inserted inside carbon nanotubes are generally well known in their stable phase. But what happened when these same molecules are confined inside a carbon nanotube? Furthermore, changes in the electronic and mechanical properties of carbon nanotubes induced by the insertion of these molecules have been demonstrated [9, 10]. Peapods are typically characterized by one or more of the conventional techniques such as transmission electron microscopy (TEM), Raman spectroscopy, electron diffraction, electron energy loss spectroscopy (EELS), and X-ray diffraction. Raman spectroscopy is a useful tool to characterize carbon nanotubes and related nanomaterials and widely used by experimentalists as a fast and nondestructive method to identify the type of nanoparticle and to study their electronic and vibrational properties [11]. In this chapter, the structure and vibrational properties of C_{60} and C_{70} peapods are reviewed. We show that the structure of the

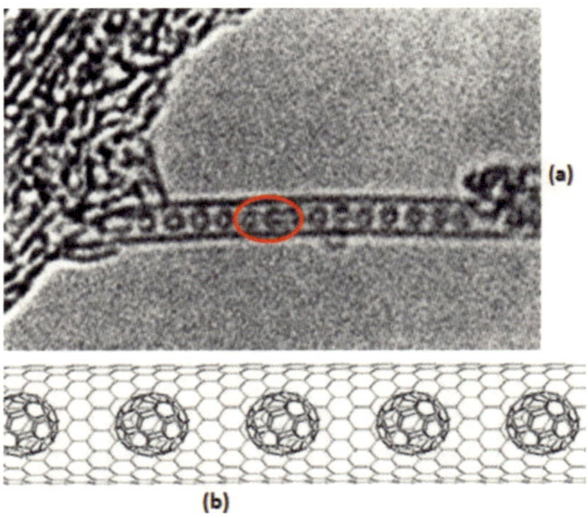

Figure 1. (a) Electron microscopy image of C_{60} peapod (from reference [4]). (b) Schematic representation of the molecular structure of an individual C_{60} peapod.

C_{60} and C_{70} molecules encapsulated inside SWCNTs adopt different configurations according to the nanotube diameter. Then, we report for each structure its associated calculated Raman responses. The dependencies of the Raman spectrum with different structural parameters such as nanotube diameter, fullerenes configuration, and the filling level are discussed. Finally, we evaluate, for each peapod configuration, the reliability and the transferability of the experimental method proposed by Kuzmany [12] to estimate the relative C_{60} and C_{70} concentrations in peapods.

2. Structure and dynamics of C_{60} and C_{70} peapods

2.1. Configuration of C_{60} and C_{70} inside carbon nanotubes

A C_{60} (C_{70}) carbon peapod is considered to be consisting of a number of C_{60} (C_{70}) molecules inside SWCNT. The configurations of C_{60} and C_{70} molecules within nanotube are diameter dependent. Carbon peapods can be produced in a very high yield close to 90% simultaneously by heating opened SWCNTs and fullerenes in a scaled quartz tube [6] and by the vapor phase reaction by using open end SWCNTs [7, 8]. The encapsulated molecules peas and the SWCNT pod are bonded through van der Waals interactions [6].

2.1.1. C_{60} peapods

Several theoretical and experimental studies have been reported on C_{60} formed within SWCNTs, and several interesting structural properties have been predicted or observed. In particular, theoretical calculations of C_{60} peapods suggest that the smallest tube diameter for encasing C_{60} molecules inside SWCNT is around the diameter corresponding to (10,10) or (9,9) SWCNTs [7]. Hodak and Girifalco showed that the guest molecules structure within the nanotube is diameter dependent [13, 14]. Using a convenient Lennard-Jones expression of the van der Waals intermolecular potential to derive the optimum configurations of C_{60} molecules inside single wall carbon nanotubes, Chadli et al. have found that the C_{60} molecules adopt a linear configuration with SWCNT diameters below 1.45 nm and a zigzag configuration for SWCNT diameters between 1.45 and 2.20 nm [15–17] (see **Table 1**). The optimum C_{60} packing can be characterized by the angle formed by three consecutive C_{60} (see **Figure 2a**). This angle θ is found to depend primarily on the nanotube diameter and does not depend significantly on the nanotube chirality. In the following paragraphs, the peapods in which the C_{60} molecules adopt a linear (zigzag) configuration are called linear (zigzag) peapods.

The calculations of structural parameters of C_{60} peapods are extended to a larger range of nanotube diameters in which C_{60} molecules can adopt a double helix (**Figure 2b**) or a two-molecule layer (**Figure 2c**) configuration. When the tube diameter increases up to 2.28 nm, the energy minimizations show that two other optimal configurations of C_{60} molecules are possible: a double helix structure ($2.15 \leq D \leq 2.23$ nm) and a two-molecule layer ($2.23 \leq D \leq 2.28$ nm). Optimized structural parameters issued from the energy minimizations are listed in **Table 2**.

Tube index (n,m)	Diameter (nm)	C60-tube distance (nm)	Angle θ (deg)	C60-C60 distance (nm)
(10,10)	1.36	0.323	180	1.003
(15,4)	1.41	0.355	180	1.003
(11,11)	1.49	0.321	164	1.006
(18,4)	1.59	0.309	150	1.003
(12,12)	1.63	0.306	145	1.008
(21,0)	1.64	0.307	143	1.008
(15,10)	1.70	0.30	134	1.006
(22,0)	1.72	0.302	132	1.006
(13,13)	1.76	0.301	127	1.007
(23,0)	1.80	0.30	121	1.003
(18,9)	1.86	0.298	112	1.003
(14,14)	1.90	0.30	108	1.006
(25,0)	1.96	0.298	99	1.006
(19,10)	2.00	0.298	90	1.008
(15,15)	2.03	0.3025	88	1.004
(17,14)	2.11	0.296	70	0.998

Table 1. Optimized linear and zigzag structural parameters of the C_{60} molecules inside SCNTs for different nanotubes diameter and chirality.

The optimal interlayer C_{60}-SWCNT distance is calculated around 0.30–0.33 nm, which is close to gaps commonly observed in carbon systems. For all optimized configurations, the interfullerene C_{60}-C_{60} distance varies from 0.998 to 1.01 nm. This result is in good agreement with the previously reported peapod interball separation of 0.97 nm from electron-diffraction profiles [18] and 0.95 nm from HRTEM data [7]. The predicted phases of C_{60}s inside SWCNTs have been experimentally observed by Kholoystov et al. HRTEM micrographs [19].

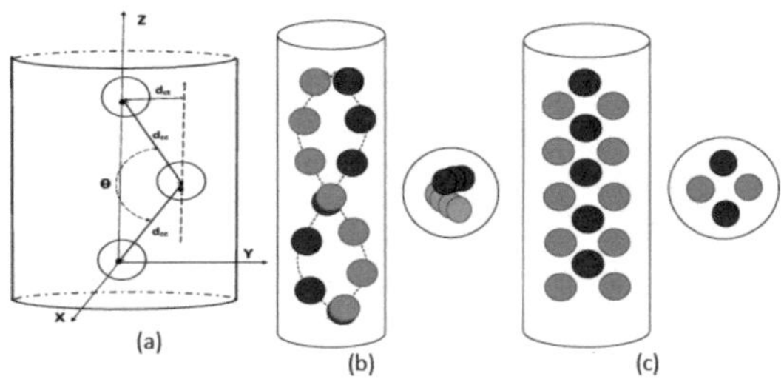

Figure 2. (a) Schematical representation of carbon peapods showing some of the parameters used for the geometrical optimization of the C_{60} molecules inside the nanotube (see text). (b) and (c) Schematic view of ordered phases resulting from C_{60} packing in SWCNTs: (b) double helix and (c) two-molecule layers.

http://dx.doi.org/10.5772/intechopen.71246

C60 phases	Tube index (n,m)	Tube diameter (nm)	C60: tube distance (nm)	C60 – C60 Distance (nm)
Double helix	(22,9)	2.164	0.321	1.003
	(16,16)	2.171	0.323	1.001
	(18,14)	2.176	0.325	0.998
	(25,5)	2.181	0.323	1.001
	(19,13)	2.183	0.329	0.998
	(28,0)	2.193	0.331	1.001
	(27,2)	2.197	0.312	1.003
	(24,7)	2.206	0.309	1.003
	(26,4)	2.210	0.308	1.006
	(22,10)	2.221	0.30	1.008
Two molecule layers	(28,1)	2.233	0.312	1.01
	(17,16)	2.239	0.297	1.002
	(18,15)	2.242	0.302	1.008
	(19,14)	2.247	0.308	1.005
	(20,13)	2.255	0.306	1.006
	(24,8)	2.259	0.307	1.008
	(21,12)	2.266	0.311	1.01
	(29,0)	2.271	0.303	1.007
	(28,2)	2.276	0.308	1.009
	(22,11)	2.280	0.310	1.01

Table 2. Packing phases parameters obtained from minimized energy of the C_{60} molecules inside SCNT with diameters between 2.16 and 2.28 nm.

2.1.2. C_{70} *peapods*

Fullerenes with ellipsoidal shape-like C_{70} are of particular interest. Unlike the spheroidal molecules such as C_{60}, there are several geometrically distinct orientations possible for the C_{70} molecule within a nanotube. Experimentally, depending on the nanotube diameter, two different orientations (with regard to the nanotube axis) of a C_{70} molecule encapsulated into SWCNTs are observed by Chorro et al.: the lying down orientation where the long axis of C_{70} molecules is parallel to the nanotube long axis, and the standing up orientation where the C_{70} long axis is perpendicular to the nanotube axis [19, 20]. The value of the nanotube diameter beyond which the change from the lying to standing orientation occurs is experimentally estimated to ~1.42 nm. Besides, HRTEM measurements showed that there is no SWCNT containing C_{70} in both orientations.

These results indicate that only one type of orientation can exist for a given nanotube and confirm that the C_{70} alignment is associated with the geometrical parameters of the nanotube host. Theoretical studies showed that the configuration of C_{70}s inside SWCNT depends primarily on tube diameter and does not significantly depend on tube chirality [21, 22]. In order to obtain the optimal structure of the C_{70} inside the nanotubes, an energy minimization procedure was performed by Fegani et al. (for more detail, see reference [21]). The authors found that the C_{70} molecules adopt a lying orientation for small SWCNT diameters (below 1.356 nm), whereas a standing orientation is preferred for large diameters (above 1.463 nm). Between these diameters, an intermediate tilted regime where C_{70} are tilted have been obtained (see **Figure 3**).

Structural parameters issued from the energy minimization are listed in **Table 3**. The optimum fullerene packing can be characterized by the inclination θ of the molecule long axis with respect to the nanotube axis: $\theta = 0°$ for lying orientation, $\theta = 90°$ for standing orientation, and $0° < \theta < 90°$ for tilted orientation.

2.2. Elements of lattice dynamics

An extensive review on phonon dynamics is out of the scope of this chapter. The lattice dynamical properties comprise the phonon modes, which provide fundamental information regarding the interatomic interaction within the material, and direct connections can be made to macroscopic phenomena and properties. The lattice dynamics theory was established by Born in the 1920s and further developed by Debye, Einstein, and others. For a comprehensive description and discussion regarding lattice dynamic theory, readers are referred to the book by Born and Huang [23]. We restrict the presentation to vibrations within the harmonic approximation. In this case, the expression of the hamiltonianis:

$$H = \frac{1}{2}\sum_{i\alpha} M_i \dot{u}_\alpha (i)^2 + \varnothing \qquad (1)$$

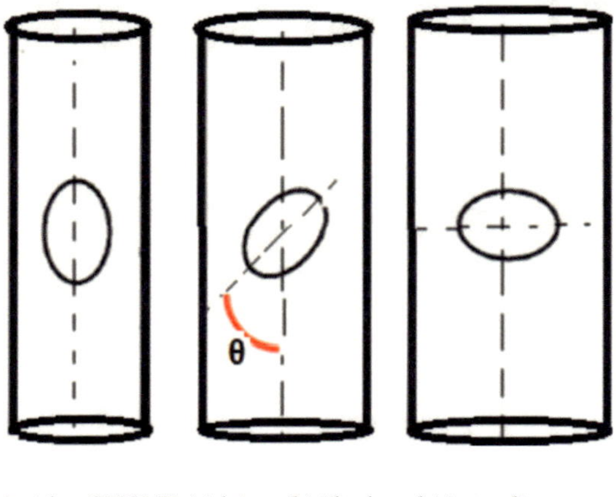

Figure 3. Orientations of a C_{70} inside a SWCNT: (a) lying, (b) tilted, and (c) standing.

SWNT index (n,m)	SWNT diameter (nm)	C_{70}-SWNT distance (nm)	Angle θ (deg)	C_{70}-C_{70} distance (nm)
(17,0)	1.330	0.307	0	1.125
(14,5)	1.335	0.309	0	1.125
(10,10)	1.356	0.320	0	1.125
(18,0)	1.409	0.307	5	1.122
(17,2)	1.415	0.305	9	1.118
(12,9)	1.428	0.307	41	1.106
(13,8)	1.437	0.309	47	1.101
(17,3)	1.463	0.323	60	1.082
(19,0)	1.487	0.329	90	1.003
(16,5)	1.487	0.329	90	1.003
(11,11)	1.491	0.331	90	1.003

Table 3. Optimized structural parameters of the C_{70} molecules inside SCNT for different diameters and chiralities.

where $\varnothing$ is the potential energy of particles in interaction, while $u_\alpha(i)$ is the displacement of atom i. It is given by:

$$\varnothing = \varnothing_0 + \frac{1}{2}\sum_{ij,\alpha\beta} \varnothing_{\alpha\beta}(i,j)\, u_\alpha(i)\, u_\beta(j) \tag{2}$$

$\varnothing_0$ is the static potential energy of the system (molecule or crystal). The starting point is the calculation of the dynamical matrix D given by:

$$D_{\alpha\beta}(i,j) = \frac{1}{\sqrt{M_i M_j}} \varnothing_{\alpha\beta}(i,j) \tag{3}$$

where $\varnothing_{\alpha\beta}(i,j)$ are the force constants between i and j atoms, the mass M_i (M_j) of the $i(j)^{th}$atom. The symmetry of the system permits to reduce the number of independent $\varnothing_{\alpha\beta}(i,j)$ coefficients.

The dynamical matrix is the key component for the computation of accurate vibrational frequencies and normal modes and contains all the information required for vibrational analysis within the harmonic approximation. In the case of C_{60} and C_{70} peapods, the dynamical matrix is calculated by block by using the coupling between the density functional theory (C_{60} and C_{70}), Saito force field (SWCNT) [24], and van der Waals potential (fullerene − SWCNT and fullerene-fullerene interactions) type:

$$V_{LJ}(r) = 4\epsilon\left[\left(\frac{\sigma}{r}\right)^{12} - \left(\frac{\sigma}{r}\right)^{6}\right] \tag{4}$$

where the depth of the potential well and the finite distance at which the interparticle potential is zero are given by $\epsilon = 2.964$ meV and $\sigma = 0.3407$ nm, respectively.

3. Nonresonant Raman spectra: Model and method

3.1. Raman scattering

Raman scattering from molecules and crystals was treated by many authors, among them Long [25] and Turell [26]. The main origin of the scattered light is considered to be an electric oscillating dipole, P, induced in the medium (molecules, amorphous materials, glasses, and crystals), by the electromagnetic incident field E. At first order, the induced dipole moment is given by

$$\mathbf{P} = \tilde{\alpha}\mathbf{E} \tag{5}$$

where $\tilde{\alpha}$ is the electronic polarizability tensor.

In a Raman experiment (**Figure 4**), a visible, or near infrared light, of frequencyω_1, wave vector k_1, polarization unit vector e_1, is incident in an isotropic medium, (ε_0 is the dielectric constant of vacuum, c is the speed of light in vacuum, and η_1 is the indice of refraction of the medium at the laser frequency).

The time-averaged power flux of the Raman-scattered light in a given direction, with a frequency between ω_f and $\omega_f + d\omega_f$ in a solid angle$d\Omega$, is related to the differential scattering cross section,

$$\frac{d^2\sigma}{d\Omega d\,\omega_d} = \frac{\hbar\omega_d^2}{8\,\pi^2 c^4}\,\omega_i\,\omega_d^3[n(\omega)+1]\,\Sigma_{\alpha\beta\gamma\lambda}\;e_{2\alpha}e_{2\beta}I_{\alpha\gamma\beta\lambda}(\omega)\,e_{1\gamma}\,e_{1\lambda'} \tag{6}$$

where

$$I_{\alpha\gamma\beta\lambda}(\omega) = \Sigma_j\; a^*_{\alpha\gamma}(j)\,a_{\beta\lambda}(j)\,\frac{1}{2\,\omega_j}\left(\delta(\omega-\omega_j)-\delta(\omega+\omega_j)\right), \tag{7}$$

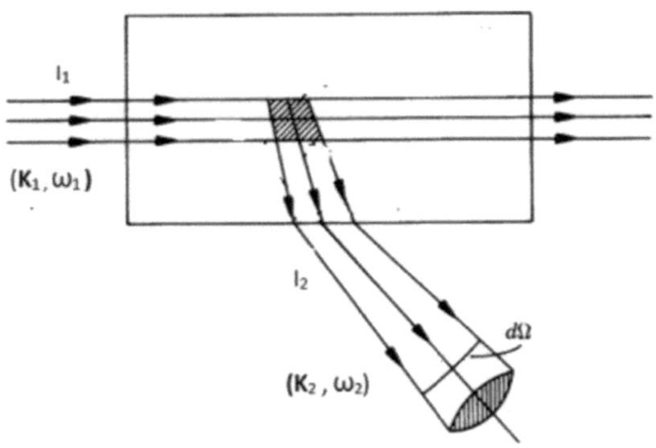

Figure 4. Sketch of a Raman scattering experiment.

http://dx.doi.org/10.5772/intechopen.71246

n(ω) is the Bose factor, the Raman shift ω is $\omega = \omega_i - \omega_d$ and

$$a_{\alpha\gamma}(j) = \Sigma_{k\delta} \frac{\pi^k_{\alpha\gamma,\delta}}{\sqrt{M_k}} e_\delta(j \mid k) \tag{8}$$

where M_k the mass of the kth atom, ω_j and $e_\delta(j \mid k)$ are the frequency and (δj) component of the jth mode. The coefficients $\pi^k_{\alpha\gamma,\delta}$ connect the polarization fluctuations to the atomic motions. They are obtained by expanding the atomic polarizability tensor $\breve{\pi}^k$ in terms of atom displacementsu^k_δ, with

$$\pi^k_{\alpha\gamma,\delta} = \Sigma_{k'} \left(\frac{\partial \pi^k_{\alpha\gamma}}{\partial u_\delta(k)}\right)_0 \tag{9}$$

$a_{\alpha\gamma}(j)$ is the $\alpha\gamma$ component of the so-called Raman polarizability tensor of the jth mode, and ω_j is the frequency of the jth vibration mode.

3.2. The bond polarizability model

The Raman intensities can be calculated within the framework of the nonresonant bond polarizability model [27]. The basic assumption of the bond polarizability model is that the optical dielectric susceptibility of the material or molecule can be decomposed into individual contributions, arising only from the polarizability α^b_{ij} of bonds b between nearest neighbor atoms. In this model, each bond is characterized by a longitudinal polarizability, α_l, and a polarizability perpendicular to the bond, α_p. Thus, the polarizability contribution of a particular bond b can be written as follows:

$$\pi^b_{ij}(r) = \frac{1}{3}\left(\alpha_l + 2\,\alpha_p\right)\delta_{ij} + \left(\alpha_l - \alpha_p\right)\left(\boldsymbol{r}_i \mathbf{r}_j - \frac{1}{3}\delta_{ij}\right), \tag{10}$$

where i and j are the Cartesian directions (x, y, z) and $\boldsymbol{r}$ is the unit vector along the bond b which connects the atom n and the atom m covalently bonded. Within this approach, one can assume that the bond polarizability parameters are functions of the bond lengths r only. The derivatives of Eq. (10) with respect to the atomic displamcent of the atom n in the direction k, $\pi^n_{ij,k}$, are linked to the Raman susceptibility of modes (see [27] for the detailed formalism) and are given by:

$$\pi^n_{ij,k} = \Sigma_m \frac{1}{3}\left(\alpha'_l + 2\alpha'_p\right)\delta_{ij}\,\boldsymbol{r}_k + \left(\alpha'_l - \alpha'_p\right)\left(\boldsymbol{r}_i \mathbf{r}_j - \frac{1}{3}\delta_{ij}\right)\boldsymbol{r}_k + \frac{\alpha_l - \alpha_p}{r}\left(\delta_{ik}\mathbf{r}_j + \delta_{jk}\boldsymbol{r}_i - 2\,\boldsymbol{r}_i \mathbf{r}_j \mathbf{r}_k\right) \tag{11}$$

where $\alpha'_l = \left(\frac{\partial \alpha_l}{\partial r}\right)_{r=r0}$, $\alpha'_p = \left(\frac{\partial \alpha_p}{\partial r}\right)_{r=r0}$ and r_0 is the equilibrium bond distance. The values of these parameters (α') are usually fitted with respect to the experiments. For example, in the case of the C_{60}, the sum over bonds in Eq. 11 includes two types of bonds, single and double, so there is one parameter that determines the Raman intensities for SWCNT.

3.3. The spectral moment's method

The direct method to calculate the Raman spectrum requires, besides the polarization parameters, direct diagonalization of the dynamical matrix to obtain the eigenvalues and the eigenvectors of the system. The diagonalization fails or requires long computing time when the system contains a large number of atoms, as for a long C_{60} and C_{70} chains inside nanotubes. By contrast, the spectral moment's method allows computing directly the Raman responses of harmonic systems without any diagonalization of the dynamical matrix [28, 29]. In the case of C_{60} and C_{70} peapods, calculations of the Raman spectra of peapods showed that approximately 500 moments are sufficient to obtain good results for larger samples (~25,000 degrees of freedom).

4. Raman spectroscopy of C_{60} and C_{70} peapods

4.1. Raman-active modes in C_{60} and C_{70} fullerene

Because of their structural and vibrational properties, a lot of theoretical and experimental studies have been presented on C_{60} and C_{70} fullerenes, and several interesting properties have been predicted or observed. In particular, experimentalists have reported Raman scattering spectra of C_{60} and C_{70} measured by depositing fullerenes on the electrode or metal surface in order to obtain more intensive and more distinguishable spectral signals [30–32]. The free C_{60} and C_{70} cluster belongs to the Ih and D5h symmetry group, respectively. According to group theory [33, 34], the high symmetry of C_{60} gives rise to 10 Raman-active modes (8Hg ⊕ 2Ag), while C_{70} has 53 Raman-active vibrational modes decomposed (12A1′ ⊕ 22E2′ ⊕ 19E1″).

A large variety of theoretical methods have been applied to the calculation of C_{60} and C_{70} vibration of the internal modes and to the determination of their Raman activity. These approaches can be mainly classified as force field model [35, 36], modified neglect of diatomic overlap (MNDO) [37], Quantum-mechanical Consistent Force Field Method for Pi-Electron Systems (QCFF/PI) [38], and density functional theory [39]. The experimental and calculated frequency (cm^{-1}) of Raman-active modes in free C_{70} molecule with their assignments is listed in **Table 4**.

4.2. Raman spectra of carbon nanotubes

Raman spectroscopy is one of the primary methods used to yield the geometrical structure of one isolated individual SWCNT [40] and organized into bundles [11]. The experimental Raman spectra of SWCNTs are dominated by the radial breathing mode (RBM), the D-band, the G-band, and the G′-band (see **Figure 1** in [40]).

The fundamental two Raman-active modes are the RBM below 350 cm^{-1} and the tangential mode (TM) located between 1400 and 1600 cm^{-1}. The RBM is an important mode for the characterization and identification of specific nanotubes, in particular of their chirality. Recently, experimental and theoretical Raman studies have shown that the RBM frequency

Experiment				**Calculation**		**Assignment**
Ref [53]	Ref [54]	Ref [55]	Ref [56]	Ref [57]	Ref [21]	
226	228		233	219	218	E_2'
251			250	245	243	E1″
259	258	261	257	253	252	A1′
396	399	400	396	393	391	A1′
410	409	411	410	408	409	E1″
455	455	459	452	448	451	A1′
506	508	501	506	502	503	E2′
568	569	573	569	564	565	A1′
701	701	704	701	701	704	A1′
713	714		714	713	708	E1″
736	737	739	738	734	739	E2′
768	766	770	769	750	749	E2′
1060	1062	1062	1060	1060	1067	A1′
1182	1182	1186		1185	1186	A1′
				1186	1193	E2′
1227	1227	1231	1227	1229	1237	A1′
1257	1257	1260	1256	1256	1253	E2′
1296	1296	1298	1296	1296	1298	E1″
	1313	1317	1311	1312	1306	E1″
1331	1335	1336	1333	1328	1325	E2′
1367	1370	1370	1366	1366	1359	E1″
1467	1469	1471	1468	1471	1473	A1′
1494	1493			1501	1501	E2′
1511	1515	1517	1512	1515	1512	E1″
1564	1565	1569	1566	1574	1583	A1′

Table 4. Experimental and calculated frequency (cm^{-1}) of Raman-active modes in C_{70} fullerene with their assignments.

vs. diameter (d) follow the phenomological law $\omega_{RBM} = \left(\frac{228}{d}\right)\sqrt{1+Cd^2}$ with $C=0$ for isolated SWCNTs [65], and $C\neq0$ for SWCNT packed in bundles (see [41] and references therein). In SWCNT bundles, the origin of the C term results from van der Waals interaction. The G-band which closely related to vibrations in all sp2 carbon materials is an intrinsic feature of a carbon nanotube. The Raman line shape of this band which depends on whether the nanotube is semiconducting or metallic allows distinguishing between both types. This band shows a low- and high-frequency component associated to the vibrations along the circumferential direction (G−) and the G direction of the nanotube axis (G+), respectively. In both metallic and semiconducting nanotubes, the former component is found to be dependent on the diameter of a nanotube, while the latter does not exhibit this dependence [40]. In the Raman spectra of isolated semiconducting and metallic SWCNT, the D-band and G′-band features are both observed.

4.3. Raman experiments on C_{60} and C_{70} peapods

Raman spectroscopy was employed to further characterize the peapod samples and to obtain structural information on the inserted fullerene inside tubes. Many groups have performed the Raman experiments of filling SWCNTs with C_{60} or C_{70} molecules [12, 42–46] since the first synthesized by Smith et al. [4]. **Figure 5** presents two examples of peapod experimental Raman spectra, one for C_{60}@SWCNT and the other for C_{70}@SWCNT.

Usually, experimental Raman spectra have been obtained on an ensemble of peapods organized into bundles. Nevertheless, the analysis of these experimental results for bundles of C_{60} peapods [8, 44] leads to the well-established conclusion: the Peapod Radial Breathing-Like Mode (PRBLM) of the peapod having a tube diameter close to 1.37 nm was split into two components, downshifted and upshifted, respectively, with respect to the position of the RBM in empty SWCNT. The upshift was possibly assigned to the stress feeling by the tubes from the inside C_{60}. For diameters between 1.45 and 1.76 nm, a single PRBLM measured

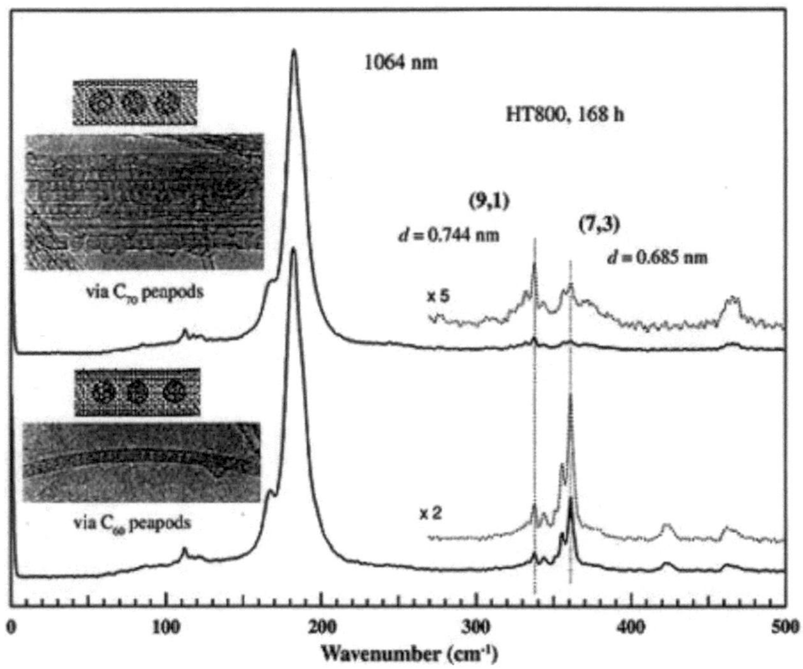

Figure 5. RBM Raman spectra taken for heat-treated C_{70} (top) and C_{60} (bottom) peapods (from reference [46]).

http://dx.doi.org/10.5772/intechopen.71246

and its frequency are downshifted compared to the RBM in the empty SWCNT bundles. The observed downshift of the PRBLM [43, 44] is explained by the hybridization effect between SWCNTs and C_{60} molecules electronic states, leading to a decreasing of the electron density in the vicinity of the SWCNT, which induces a decreasing of the force constant of the C-C bond. Pfeiffer et al. [45] measured all the fundamental Raman lines of the encaged C_{60} peas. They observed that both nondegenerate and totally symmetric Ag modes of C_{60} peas exhibit a splitting into two components. They attributed this splitting to the presence of both moving and static fullerenes inside the tubes.

Kuzmany et al. [12] performed a detailed Raman analysis to evaluate the concentration of C_{60} molecules inside nanotubes. As expected, the relative concentrations derived from the measurement of normalized intensity ratio for each Raman mode of C_{60} are close. The C_{60} filling degree of a reference peapod sample was determined from electron energy loss spectroscopy experiment in order to evaluate the absolute C_{60} concentration of each peapod sample from the measurements of the Raman spectra (see **Table 1** of [12]).

Concerning the C_{70} peapods, Hirahara et al. [46] characterized one-dimensional crystals of a variety of larger fullerenes C_{70} peapods by using high-resolution transmission electron microscopy and electron diffraction. They concluded that the C_{70} admit two different orientations depending on the nanotube diameter: the lying and standing orientation. The intermolecular distances of various fullerenes in SWCNTs are considerably smaller than those for bulk fullerene crystals, suggesting an effect of confinement in the one-dimensional channels inside SWCNTs. Raman experiments on SWCNTs encasing C_{70} molecules have been reported [19, 21, 42, 47, 48]. Ryabenko et al. [42] concluded that the PRBLM mode of C_{70} peapods are downshifted by ~2–3 cm^{-1} compared to empty nanotubes. For different orientations of C_{70}s in C_{70}@SWCNT peapods organized into bundles, the measured PRBLM downshift after the C_{70} encapsulation is ~2–6 cm^{-1} [47]. The downshift of the PRBLM suggests a structural relaxation or the tube diameter transformation is expected to occur with the assistance of such added carbon atoms. However, such experimental work should include Raman investigations on samples showing peapods having various structural characteristics: different tube diameters, different fullerene concentration, and bundles with various sizes, for example, various numbers of tubes.

4.4. Raman-active modes calculation in C_{60} peapods

The Raman cross section was calculated assuming that scattering can be described within the framework of the bond polarizability model [52]. Recently, we calculated the Raman spectra of infinite homogeneous bundles (crystal) of C_{60} peapods [49]. In order to reach a 100% factor filling, 20 C_{60} molecules are located in a length of 19.8 nm [17.7 nm] of a (10,10) [(13,13)] nanotube and the number of carbon atoms of the tube (10,10) [(13,13)] is close of 3240 [3744] atoms. For the C_{60}@(10,10) (C_{60}@(13,13)) peapod, a 100% factor filling corresponds to a concentration of 37% (32%) (concentration is the ratio between carbons in C_{60} and carbons in the tube. In **Figure 6b** are displayed the calculated polarized ZZ Raman spectra of individual C_{60} peapods and crystal of C_{60} peapods, respectively, for linear and zigzag configuration of the C_{60} molecules inside the tube: (bottom) linear chain of

C_{60} confined into a (10,10) SWCNT (tube diameter close to 1.36 nm), (top) zigzag chain of C_{60} confined into a (13,13) SWCNT (tube diameter close to 1.76 nm).

In the intermediate and high-frequency regions, the calculated Raman spectra of individual peapods organized into infinite bundles do not show significant differences. It can be emphasized that a splitting of the Ag(1) and Ag(2) totally symmetric C_{60} modes was observed experimentally [45]. This splitting was considered as signature of the mobility of some encapsulated C_{60} molecules inside nanotubes [45]. In our calculations, the C_{60} molecules are kept at fixed positions in the considered peapods, and no splitting of the totally symmetric C_{60} modes is found.

Next, we show in **Figure 6b** the low-frequency range of Raman spectra of C_{60}@(10,10) and C_{60}@(13,13) peapods where the main changes are identified. We found that the PRBLM of individual C_{60}@(10,10) linear peapod on one hand and the PRBLM(1) and PRBLM(2) of individual C_{60}@(13,13) zigzag peapod on the other hand are slightly upshifted in bundles of C_{60}@(10,10) peapods and C_{60}@(13,13) peapods, respectively. We calculated a shift of the RBLM from 170 cm^{-1} in the individual C_{60}@(10,10) peapod to 186 cm^{-1} in the infinite bundle of C_{60}@(10,10) peapods. In the case of C_{60}@(13,13), the PRBLM(1) and PRBLM(2) doublet calculated

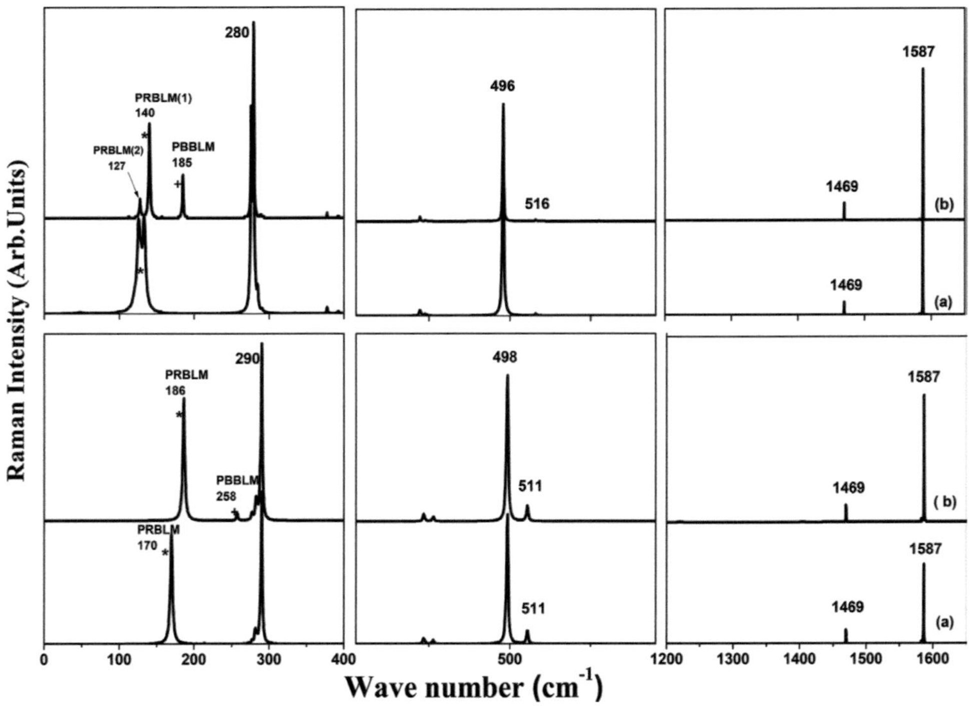

Figure 6. The ZZ calculated Raman spectra of individual (curve (a)) and homogeneous bundles (curve (b)) of C_{60}@(13,13) (top) and C_{60}@(10,10) (bottom) (filling factor 100%). The stars give the positions of the RBM and RBLM in individual SWCNT and crystal of SWCNT, respectively. Cross indicates the frequency of Bundle Breathing Like-Mode (BBLM) in crystal of SWCNT. (left) Breathing modes range, (middle) region of the radial C_{60} modes, (right) range of tangential modes: Ag(2) mode of C_{60} and G-modes of the nanotubes.

for the individual C_{60}@(13,13) peapod at 133 and 125 cm^{-1}, respectively, shifts to 140 and 127 cm^{-1} in the infinite bundle of C_{60}@(13,13) peapod. A new peapod bundle breathing-like mode (PBBLM) is also calculated in the Raman spectra of an infinite bundle of C_{60} peapods, which arises from the BBLM in SWCNT bundle [50]. Concerning the main Raman-active modes of C_{60} (e.g., modes located around 270 cm^{-1} (Hg) and 495 cm^{-1} (Ag)), independently of the configuration of C_{60} molecules inside the tube, the van der Waals interactions between peapods have no significant effect on these modes.

4.4.1. Diameter dependence of the Raman spectrum of C_{60} peapods

Figure 7 shows the PRBLM and PBBLM frequency dependencies as a function of the tube diameter. These dependencies are first compared with those of the RBLM and BBLM in unfilled bundles of tubes. The behavior of the PRBLM in linear peapods is qualitatively the same as those of RBLM in SWCNT bundle. However, the PRBLM frequency versus diameter relation deviates from the scaling law stated for the RBLM frequency in SWCNT bundle, especially for small diameters. For zigzag peapod, the PRBM(1) dependence as a function of the diameter is close to that of the RBLM in bundle of SWCNT. The PRBLM(1) downshifts when the tube diameter increases. Concerning the PBBLM frequency, it decreases with increasing the tube diameter and close to the frequency of BBLM, except at small diameters where a more significant increase of the PBBLM frequency is calculated. Our calculations based on DFT have clarified that the shift in the RBM frequencies of nanotubes containing fullerenes strongly depends on the diameters of the nanotubes. DFT calculations have also clarified that the shift in the RBM frequencies of nanotubes containing fullerenes strongly depends on the diameters of the nanotubes [51, 53–57].

More recently, Raman calculations are extended to a larger range of diameters in which C_{60} molecules can adopt a double helix and a layer of two molecules [22].

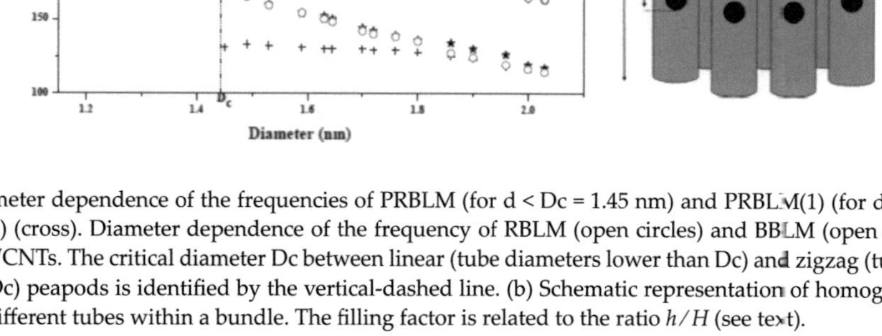

Figure 7. Diameter dependence of the frequencies of PRBLM (for d < Dc = 1.45 nm) and PRBLM(1) (for d > Dc) (stars), and PRBLM(2) (cross). Diameter dependence of the frequency of RBLM (open circles) and BBLM (open diamonds) in bundles of SWCNTs. The critical diameter Dc between linear (tube diameters lower than Dc) and zigzag (tube diameters greater than Dc) peapods is identified by the vertical-dashed line. (b) Schematic representation of homogeneous filling mode of the different tubes within a bundle. The filling factor is related to the ratio h/H (see text).

4.4.2. *Filling rate effect*

In real carbon peapod samples, it is reasonable to consider that all the nanotubes are not completely filled with fullerenes. The degree of filling ranges from a certain percent to almost 100% [6]. In the following paragraphs, we discuss the main features of the filling rate effect on both configurations of C_{60} inside SWCNT, double helix, and two-molecule layer. We assume that the molecules tend to cluster inside nanotubes. Indeed, this should correspond to a low energy configuration of the system as the energy is lowered by the attractive C_{60}-C_{60} interactions. The filling factor is defined as the number of carbon atoms of C_{60} molecules contained in the h length with regard to the number of carbon atoms contained in the H length of the host tube normalized with the concentration related to the filling factor of 100% (see **Figure 7b**). The calculated ZZ-polarized Raman spectra of the C_{60}@(28,0) (double helix chain of C_{60}) and C_{60}@(29,0) (two-molecule layer) peapods are reported in **Figure 8** as a function of five filling rates (20, 40, 60, 80, and 100%). Where the RBLM range is very sensitive to filling rate and the TLM range slightly depends on the degree of filling of the SWCNT, the TLM range is not shown.

For both configurations, a single PRBLM is predicted whatever the tube filling. For the empty tubes, the RBM is located at 102 and 98 cm^{-1} for the (28,0) and (29,0) SWCNTs, respectively. For a high filling level, it is upshifted at 103 and 100 cm^{-1} in the C_{60}@(28,0) and C_{60}@(29,0) peapods, respectively. As expected, the intensity of the Hg(1) line located at 270 cm^{-1} in C_{60} increases when the filling factor increases.

4.5. Calculation of Raman-active modes in C_{70} peapods

In this section, we review the theoretical Raman spectra of the C_{70} peapods calculated as a function of diameter and C_{70} filling rate. These calculations support the experimental

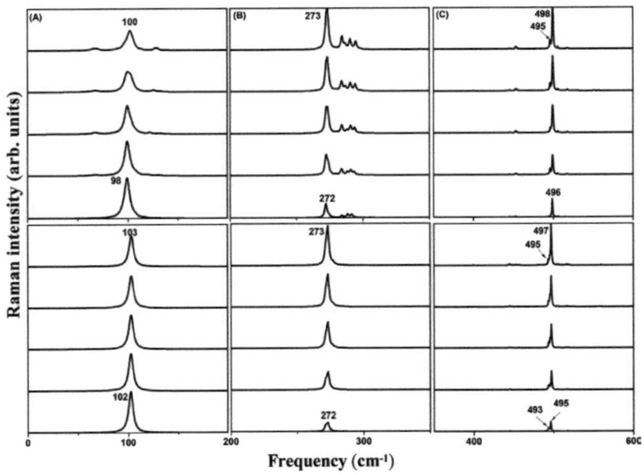

Figure 8. Calculated ZZ-polarized Raman spectra of C_{60}@(28,0) (bottom) and C_{60}@(29,0) (top) as a function of the filling rate and displayed in the BLM region. From bottom to top, the filling rate is: 20, 40, 60, 80, and 100%.

evidence on the C_{70} orientations as a function of the SWCNT diameters, and they address new questions as to the influence of the nanotube chirality and the C_{70} filling rate.

4.5.1. Diameter effects

In order to investigate how the frequency of the Raman-active mode in C_{70}@SWCNT changes when the C_{70} molecule adopts various orientations, and three zigzag SWCNTs have been considered with a diameter of 1.330 [(17,0)], 1.409 [(18,0)], and 1.487 nm [(19,0)] where C_{70} molecules adopt lying, tilted, and standing orientation, respectively. The calculated ZZ-polarized Raman spectra of C_{70} peapods are shown in **Figure 9** along with their corresponding unfilled nanotubes and the unoriented C_{70} molecule. Raman lines can be divided into three frequency ranges: (1) below 500 cm^{-1} where the breathing-like modes (BLM) dominate (panel A), (2) an intermediate range between 500 and 1500 cm^{-1} (panel B), and (3) above 1500 cm^{-1} where the tangential-like modes (TLM) are located (panel C).

In the TLM range, the main modes of SWCNT are almost not significantly sensitive to the orientation of C_{70} molecules inside the nanotube. In the intermediate range, the C_{70} spectrum is dominated by two strong lines at 1192 and 1253 cm^{-1}, while no Raman line is expected for SWCNTs. Thus, Raman spectra of peapods show several weak lines which originate from the splitting of the C_{70} degenerate modes due to van der Waals interactions. In this range, Raman spectra of peapods are dominated by two lines around 740 and 1272 cm^{-1}. A third line can also be identified at 1391 cm^{-1} for the standing orientation and at 1490 cm^{-1} for lying and tilted orientations.

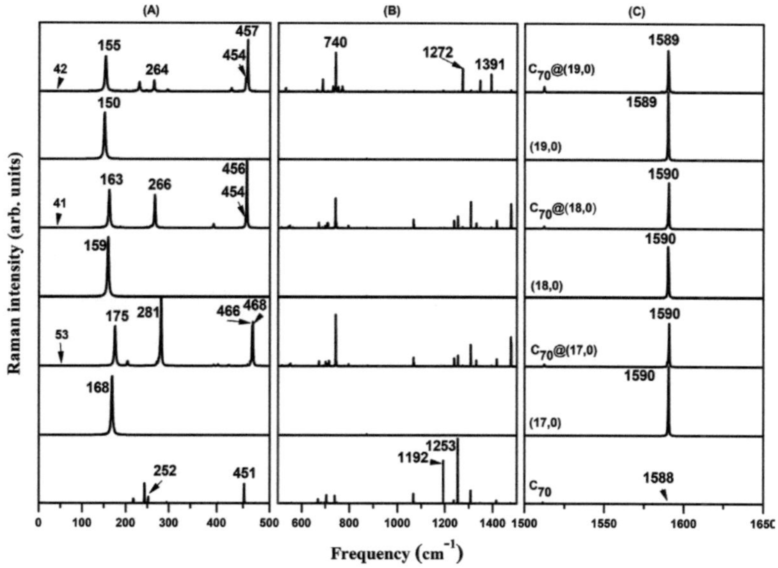

Figure 9. ZZ-calculated Raman spectra of C_{70} peapods along with their corresponding unfilled nanotubes and unoriented free C70 molecule. Spectra are displayed in the BLM (A), intermediate (B), and TLM (C) ranges.

4.5.2. *Filling factor effects*

To investigate the effect of incomplete filling rate on Raman spectra of C_{70} peapod, we considered three kinds of peapods: the C_{70}@(17,0) lying, C_{70}@(17,3) tilted, and C_{70}@(11,11) standing orientations. The spectra were calculated for five values of the filling factor F = 20, 40, 60, 80, and 100% corresponding to 2, 4, 8, 12, and 16 C_{70} molecules inside SWCNTs, respectively (see **Figure 10**). Periodic conditions were applied along the tube axis to avoid finite size effects.

As expected, the filling level of C_{70} molecules inside the SWCNTs has no significant effect on the Raman spectrum in the TLM range (not shown). In contrast, in the BLM range, the Raman spectra are quite sensitive to the filling factor: an upshift of approximately 2–5 cm^{-1} is obtained for the RBM modes in standing orientations when F increases up to 100%, whereas a splitting of these modes is observed in the lying and tilted orientations. This splitting is in agreement with experimental results of Bandow et al. [8]. For example, for lying orientation (small diameters), the increase of the filling factor from 0 to 100% leads to the appearance of two lines at frequencies close to 168 and 175 cm^{-1}. Their intensity shifts from the one located around 168 cm^{-1} to that located around 175 cm^{-1} as the filling factor increases.

4.6. Raman spectra analysis: C_{60} and C_{70} concentration determination in peapods

The evolution of the nonresonant Raman scattering intensity ratios between the Raman mode of fullerene peas and nanotubes as a function of the fullerene concentration inside the tubes has been investigated in the framework of the spectral moment's method [21, 50]. Although the nonresonant approach cannot predict the variation of the line intensities with the excitation energies, the obtained predictions are useful to follow their evolution as a function of the filling rate of fullerenes inside SWCNT. For this purpose, we performed an average of

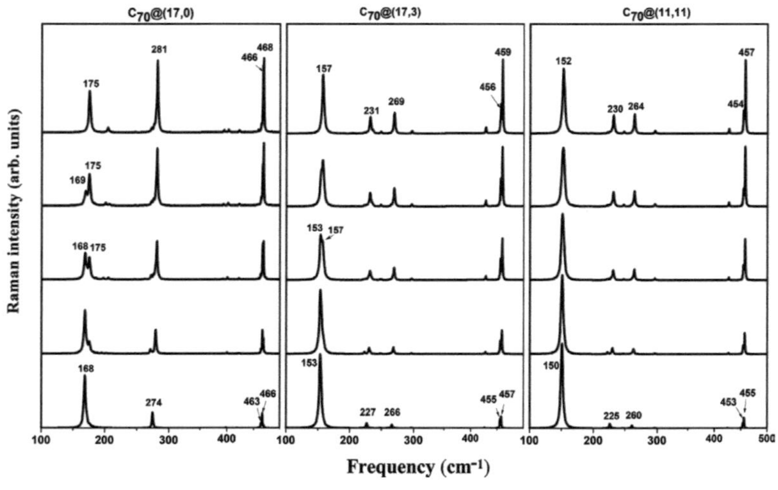

Figure 10. ZZ-polarized Raman spectra of infinite peapods displayed in the BLM region as a function of the filling factor: F = 20, 40, 60, 80, and 100% from bottom to up.

the Raman spectra over the peapod orientations with regard to the laboratory frame. Raman spectra are calculated in the VV configuration for unoriented peapod samples for various values of the fullerene filling factor. The relative intensity ratios have been calculated as a function of the C_{60} and C_{70} fullerenes concentrations.

First, for C_{60} peapod, we calculated the intensity ratio between the Raman lines of the C_{60} molecule [Hg(2), Ag(1), Hg(3), Hg(4), Hg(7), and Ag(2)] and the PRBLM or G-mode. In order to make the comparison with the experimental results of Kuzmany et al. [12], the calculated intensity ratio is normalized with respect to the same intensity ratio calculated for the 60% filling factor sample. The relative C_{60} concentrations that are derived according to this procedure are displayed in **Figure 11** for the C_{60}@(17,0), C_{60}@(22,0), C_{60}@(28,0), and C_{60}@(29,0) which correspond to linear, zigzag, double-helix, and two molecule layers chain of C_{60}, respectively. As expected, for all the investigated peapod diameters, the relative concentrations calculated for each C_{60} mode are close. The relative concentrations calculated for infinite peapods increase when its diameter increases. For instance, for a filling factor ~20%, the relative concentration is close to 0.21, 0.26, 0.3, and 0.4 for a diameter of 1.35 [C_{60}@(17,0)], 1.76 [C_{60}@(22,0)], 2.19 [C_{60}@(28,0)], and 2.27 nm [C_{60}@(29,0)], respectively. For a ~ 20% concentration (corresponding to the L43 sample (EELS concentration 25 ± 10) in [12]), the average relative concentration is

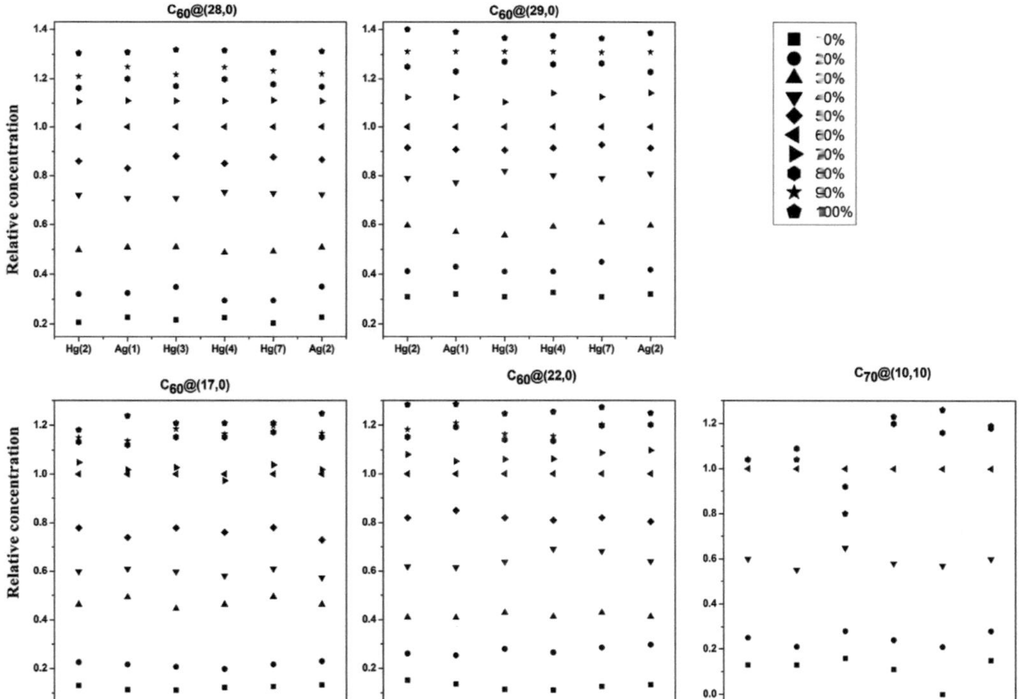

Figure 11. Relative concentration of several configurations of C_{60} inside SWCNTs and C_{70}@(10,10) normalized on the 60% filling rate intensities.

calculated around 0.21 ± 0.02, in good agreement with the experimental relative concentration evaluated around 0.19. This good agreement supports the experimental method proposed by Kuzmany et al. to evaluate C_{60} concentration inside SWCNTs.

Next, we calculate the integrated intensity ratios between the Raman lines of the C_{70} molecule [low-frequency modes located at: A1′: 252 cm^{-1} (A1′(1), A1′: 391 cm^{-1} (A1′(2)), A1′: 451 cm^{-1} (A1′(3)) and high-frequency Raman modes: A1′: 1473 cm^{-1} (A1′(4)), E1″: 1512 cm^{-1} (E1″(1)) and A1′: 1583 cm^{-1} (A1′(5))] and the PRBLM or G-mode. The relative concentrations derived by this way are shown in **Figure 11**. For a filling factor 20%, the relative concentration is close to 0.21 for C_{70} inside (10,10). The comparison of data of the relative concentration of C_{60} (linear chain) with those of C_{70} (lying orientation) shows qualitatively the same result.

5. Concluding remarks

In this chapter, we have discussed the optimal configurations of the C_{60} and C_{70} molecules inside SWCNTs. First, the configuration of C_{60} and C_{70} fullerenes inside SWCNT is found to depend strongly on the nanotube diameter and does not depend significantly on tube chirality. In agreement with experiments, the C_{70} molecules adopt a lying orientation for small SWCNT diameters (below 1.356 nm), whereas a standing orientation is preferred for large diameters (above 1.463 nm). Between these diameters, an intermediate tilted configuration of C_{70}s is found. For C_{60} peapods, C_{60}s adopt a linear arrangement for SWCNT diameter lower than 1.45 nm and a zigzag configuration for diameters between 1.45 and 2.15 nm. In the diameter range between 2.15 and 2.23 nm, C_{60} molecules adopt a double helix arrangement in SWCNTs, whereas a layer of two molecules is preferred for diameters between 2.23 and 2.28 nm. Next, we have reported the calculated Raman spectra of peapods. The modes located in the BLM range are very sensitive to the encapsulation of fullerenes because major changes in frequencies and intensities are observed. The average intensity ratios between C_{60} and C_{70} Raman-active modes and the nanotube ones, as a function of the concentration molecules, have been analyzed, and a general good agreement is found between calculations and measurements. This good agreement supports the experimental method proposed to evaluate C_{60} concentration inside SWCNTs.

Author details

Hassane Chadli[1], Fatima Fergani[2] and Abdelali Rahmani[2]*

*Address all correspondence to: a.rahmani@fs-umi.ac.ma

1 High School of Technology Khenifra - Moulay Ismail University, Meknes, Morocco

2 Laboratory of Advanced Materials Studies and Applications, Moulay Ismail University, FSM-ESTK-ESTM, Meknes, Morocco

References

[1] Kroto H, Heath J, O'Brien S, Curl R, Smalley R. Buckminsterfullerene. Nature. 1985;**318** (14):162-163. DOI: 10.1038/318162a0

[2] S. Ijima. Helical microtubules of graphitic carbon. Nature. 1991;**354**:56-58. DOI: 10.1038/354056a0

[3] Ijima S, Ichihachi T. Single-shell carbon nanotubes of 1-nm diameter. Nature. 1993;**363**: 603-605. DOI: 10.1038/363603a0

[4] Smith BW, Monthioux M, Luzzi DE. Encapsulated C_{60} in carbon nanotubes. Nature. 1998;**396**:323-324. DOI: 10.1038/24521

[5] Sloan J, Dunin-Borkowski RE, Hutchison JL, Coleman KS, Williams VC, Claridge JB, York APE, Xu C, Bailey SR, Brown G, Friedrichs S, Green MLH. The size distribution, imaging and obstructing properties of C_{60} and higher fullerenes formed within arc-grown single walled carbon nanotubes. Chemical Physics Letters. 2000;**316**:191

[6] Kataura H, Kumazawa Y, Maniwa Y, Umezu I, Suzuki S, Ohtsuka Y, Achiba Y. Highyield fullerene encapsulation in single-wall carbon nanotubes. Synthetic Metals. 2000;**121**:1195. DOI: 10.1016/S0379-6779(00)00707-4

[7] Hirahara K, Suenaga K, Bandow S, Kato H, Okazaki T, Shinohara H, Iijima S. One dimensional Metallofullerene crystal generated inside single-walled carbon nanotubes. Physical Review Letters. 2000;**85**:5384. DOI: 10.1103/PhysRevLett.85.5384

[8] Bandow S, Takizawa M, Kato H, Okazaki T, Shinohara H, Iijima S. Smallest limit of tube diameters for encasing of particular fullerenes determined by radial breathing mode Raman scattering. Chemical Physics Letters. 2001;**347**:23-28. DOI: 10.1016/S0009-2614(01)01020-X

[9] Hornbaker DJ, Kahng SJ, Misra S, Smith BW, Johnson AT, Mele ET, Luzzi DE, Yazdani A. Mapping on the one-dimensional electronic states of nanotube peapod structures. Science. 2002;**295**(5556):828-831. DOI: 10.1126/science.1068133

[10] Jaroenapibal P, Chikkannanavar SB, Luzzi DE, Evoy S. Nanomechanical resonance studies of carbon nanotube peapod bundles. The Journal of Applied Physics. 2005;**98**:044301. DOI: 10.1063/1.2001147

[11] Dressellhaus MS, Eklund PC. Phonons in carbon nanotubes. Advances in Physics. 2000;**49**(6):705-814. DOI: 10.1080/000187300413184

[12] Kuzmany H, Pfeiffer R, Kramberger C, Pichler T, Liu X, Knupfer M, Fink J, Kataura H, Achiba Y, Smith BW, Luzzi DE. Analysis of the concentration of C_{60} fullerenes in single wall carbon nanotubes. Applied Physics A: Materials Science & Processing. 2003;**76**:449. DOI: 10.1007/s00339-002-2046-8

[13] Hodak M, Girifalco LA. Ordered phases of fullerene molecules formed inside carbon nanotubes. Physical Review B. 2003;**67**:075419-075422. DOI: 10.1103/PhysRevB.67.075419

[14] Hodak M, Girifalco LA. Systems of C_{60} molecules inside (10,10) and (15,15) nanotube: A Monte Carlo study. Physical Review B. 2003;**68**:085405. DOI: 10.1103/PhysRevB.68.085405

[15] Chadli H, Rahmani A, Sbai K, Hermet P, Rols S, Sauvajol J-L. Calculation of Raman-active modes in linear and zigzag phases of fullerene peapods. Physical Review B. 2006;**74**:205412. DOI: 10.1103/PhysRevB.74.205412

[16] Chadli H, Rahmani A, Sbai K, Sauvajol J-L. Raman active modes in carbon peapods. Physica A: Statistical Mechanics and its Applications. 2005;**358**:226-236. DOI: 10.1016/j.physa.2005.06.025

[17] Chadli H, Rahmani A, Sauvajol J-L. Raman spectra of C_{60} dimer and C_{60} polymer confined inside a (10,10) single-walled carbon nanotube. Journal of Physics: Condensed Matter. 2010;**22**(14):145303. DOI: 10.1088/0953-8984/22/14/145303

[18] Liu X, Pichler T, Knupfer M, Golden MS, Fink J, Kataura H, Achiba Y, Hirahara K, Ijima S. Filling factors, structural, and electronic properties of C_{60} molecules in single-wall carbon nanotubes. Physical Review B. 2002;**65**:045419. DOI: 10.1103/PhysRevB.65.045419

[19] Khlobystov A N, Scipioni R, Nguyen-Manh D, Britz D A, Pettifor D G, Briggs G A D, Lyapin S G, Ardavan A, Nicholas R. Controlled orientation of ellipsoidal fullerene C_{70} in carbon nanotubes. The Journal of Applied Physics. 2004;**84**:792-794. DOI: 0003-6951/2004/84(5)/792/3/22.00

[20] Chorro M, Delhey A, No L, Monthioux M, Launois P. Orientation of C_{70} molecules in peapods as a function of the nanotube diameter. Physical Review B. 2007;**75**(3):035416. DOI: 10.1103/PhysRevB.75.035416

[21] Fergani F, Chadli H, Belhboub A, Hermet P, Rahmani A. Theoretical study of the Raman spectra of C_{70} fullerene carbon peapods. Journal of Physical Chemistry C. 2015;**119**:5679-5686. DOI: 10.1021/jp511969t

[22] Fergani F, Ait Abdelkader SA, Chadli H, Fakrach B, Rahmani AH, Hermet P, Rahmani A. C_{60} filling rate in carbon peapods: A nonresonant Raman spectra analysis. Journal of Nanomaterials. 2017;**2017**:1-7. DOI: 10.1155/2017/9248153

[23] Born M, Huang K. Dynamical Theory of Crystal Lattices. Oxford: Oxford University Press; 1954. p. 432. DOI: 10.1119/1.1934059

[24] Saito R, Takeya T, Kimura T, Dresselhaus G, Dresselhaus MS. Raman Intensity of Single-Wall Carbon Nanotubes. Physical Review B 545 1998, 57, 4145

[25] Long DA. The Raman Effect: A Unified Treatment of the Theory of Raman Scattering by Molecules. 2nd ed. Chichester: Wiley; 2011. p. 624. DOI: 10.1002/0470845767

[26] Turell G. Infrared and Raman spectra of crystals. London and New York: Academic Press; 1972. 384p. DOI: 10.1002/pol.1973.130110313

[27] Bilz, H., Strauch, D., Wehner, R.K. Light and Matter Id / Licht und Materie Id: Infrared and Raman Spectra of Non-Metals. vol. XXV. 1st ed. Handbuch der Physik. Berlin: Springer-Verlag; 1984, Pt. 2d, 253 p

[28] C. Benoit, E. Royer, G. Poussigue. The spectral moments method. Journal of Physics: Condensed Matter 1992;**4**:3125. DOI: 10.1088/0953-8984/4/12/010/meta

[29] Rahmani A, Sauvajol J-L, Rols S, Benoit C. Nonresonant Raman spectrum in infinite and finite single-wall carbon nanotubes. Physical Review B. 2002;**66**:125404. DOI: 10.1103/PhysRevB.66.125404

[30] Fredericks PM. Fourier transform surface-enhanced Raman scattering of C_{70} on a roughened silver surface. Chemical Physics Letters. 1996;**253**:251-256. DOI: 10.1016/0009-2614(96)00251-5

[31] Katayama N, Miyatake Y, Ozaki Y, Kikuchi K, Achiba Y, Ikemoto I, Iriyama K. Surface-enhanced Raman scattering (SERS) study of C_{60} and C_{70} in evaporated films by near-infrared FT-Raman spectroscopy. Proceedings of SPIE. 1994;**2089**:174-175. DOI: 10.1117/12.166755

[32] Luo ZX, Yan F. SERS of gold/C_{60}(/C_{70}) nano-clusters deposited on iron surface. Vibrational Spectroscopy. 2005;**39**:151-156. DOI: 10.1016/j.vibspec.2005.01.008

[33] Cotton FA. Chemical Applications of Group Theory. 3rd ed. New York: John Wiley & Sons. 1990. 461p. DOI: 10.1002/bbpc.19910950625

[34] Cyvin SJ, Brendsdal E, Cyvin BN, Brunvoll J. Molecular vibrations of footballene. Chemical Physics Letters. 1988;**143**(4):377-380. DOI: 10.1016/0009-2614(88)87050-7

[35] Jishi RA, Mirie RM, Dresselhaus MS. Force-constant model for the vibrational modes in C_{60}. Physical Review B. 1992;**45**(23):13685-13689. DOI: 10.1103/PhysRevB.45.13685

[36] Wu ZC, Jelski DA, George TF. Vibrational motions of buckminsterfullerene. Chemical Physics Letters. 1987;**137**(3):291-294. DOI: 10.1016/0009-2614(87)80221-X

[37] Stanton RE, Newton MD. Normal vibrational modes of buckminsterfullerene. The Journal of Physical Chemistry. 1988;**92**:2141-2145. DOI: 10.1021/j100319a012

[38] Negri F, Orlandi G, Zerbetto F. Quantum-chemical investigation of Franck Condon and Jahn–teller activity in the electronic spectra of buckminsterfullerene. Chemical Physics Letters. 1988;**144**(1):31-37. DOI: 10.1016/0009-2614(88)87084-2

[39] Weeks DE, Harter WG. Rotation-vibration spectra of icosahedral molecules. II. Icosahedral symmetry, vibrational eigenfrequencies, and normal modes of buckminsterfullerene. The Journal of Chemical Physics. 1989;**90**:4744-4771. DOI: 10.1063/1.456571

[40] Dresselhaus MS, Dresselhaus G, Jorio A, Souza Filho AG, Saito R. Raman spectroscopy on isolated single wall carbon nanotubes. Carbon. 2002;**40**(12):2043-2061. DOI: 10.1016/S0008-6223(02)00066-0

[41] Araujo PT, Maciel IO, Pesce PBC, Pimenta MA, Doorn SK, Qian H, Hartschuh A, Steiner M, Grigorian L, Hata K, Jorio A. Nature of the constant factor in the relation between radial breathing mode frequency and tube diameter for single-wall carbon nanotubes. Physical Review B. 2008;**77**:241403. DOI: 10.1103/PhysRevB.77.241403

[42] Ryabenko AG, Kiselev NA, Hutchison JL, Moroz TN, Bukalov SS, Mikhalitsyn LA, Loutfy RO, Moravsky AP. Spectral properties of single-walled carbon nanotubes encapsulating fullerenes. Carbon. 2007;**45**:1492-1505. DOI: 10.1016/j.carbon.2007.03.031

[43] Joung SK, Okazaki T, Kishi N, Okada S, Bandow S, Iijima S. Effect of fullerene encapsulation on radial vibrational breathing- mode frequencies of single-wall carbon nanotubes. Physical Review Letters. 2009;**103**:027403. DOI: 10.1103/PhysRevLett.103.027403

[44] Bandow S, Takizawa M, Hirahara K, Yudasaka M, Iijima S. Raman scattering study of doublewall carbon nanotubes derived from the chains of fullerenes in single wallcarbon nanotubes. Chemical Physics Letters. 2001;**337**:48-57. DOI: 10.1016/S0009-2614(01)00192-0

[45] Pfeiffer R, Kuzmany H, Pichler T, Kataura H, Achiba Y, Melle-Franco M, Zerbetto F. Electronic and mechanical coupling between guest and host in carbon peapods. Physical Review B. 2004;**69**:035404. DOI: 10.1103/PhysRevB.69.035404

[46] Bandow S, Hiraoka T, Yumura T, Hirahara K, Shinohara H, Iijima S. Raman scattering study on fullerene derived intermediates formed within single-wall carbon nanotube: From peapod to double-wall carbon nanotube. Chemical Physics Letters. 2004;**384**: 320-325. DOI: 10.1016/j.cplett.2003.12.032

[47] Guan L, Li H, Shi Z, You L, Gu Z. Standing or lying C_{70}s encapsulated in carbon nanotubes with different diameters. Solid State Communications. 2005;**133**:333-336. DOI: 10.1016/j.ssc.2004.11.008

[48] Kavan L, Dunsch L, Kataura H, Oshiyama A. Electro-chemical tuning of electronic structure of C_{60} and C_{70} fullerene peapods: In situ visible near-infrared and Raman study. The Journal of Physical Chemistry B. 2003;**107**(31):7666-7675. DOI: 10.1021/jp035332f

[49] Chadli H, Fergani F, Bentaleb M, Fakrach B, Sbai K, Rahmani A, Bantignies JL, Sauvajol JL. Influence of packing on the vibrations of homogeneous bundles of C_{60} peapods. Physica E: Low-dimensional Systems and Nanostructures. 2015;**71**:31-38. DOI: 10.1016/j.physe.2015.03.018

[50] Sbai K, Rahmani A, Chadli H, Sauvajol J-L. Raman-active modes in homogeneous and inhomogeneous bundles of single-walled carbon nanotubes. Journal of Physics: Condensed Matter. 2009;**21**(4):045302. DOI: 10.1088/0953-8984/21/4/045302

[51] Okada S. Radial-breathing mode frequencies for nanotubes encapsulating fullerenes. Chemical Physics Letters. 2007;**438**:59-62. DOI: 10.1016/j.cplett.2007.02.058

[52] Snoke DW, Cardona M. A bond polarizability model for the C_{60} Raman spectrum. Solid State Communications. 1993;**87**(2):121-126. DOI: 10.1016/0038-1098(93)90339-O

[53] Lynch K, Tanke C, Menzel F, Brockner W, Scharff P, Stumpp E. FT-Raman spectroscopic studies of C_{60} and C_{70} subsequent to chromatographic separation including solvent effects. Journal of Physical Chemistry A. 1995;**99**:7985-7992. DOI: 10.1021/j100020a022

[54] Wang KA, Zhou P, Rao AM, Eklund PC, Jishi RA, Dresselhaus MS. Intramolecular-vibrational-mode softening in alkali-metal-saturated C_{70} films. Physical Review B. 1993;**48**: 3501-3506. DOI: 10.1103/PhysRevB.48.3501

[55] Bethune DS, Meijer G, Tang WC, Rosen HJ, Golden WG, Seki H, Brown CA, Vries MS. Vibrational Raman and infrared spectra of chromatographically separated C_{60} and C_{70} fullerene clusters. Chemical Physics Letters. 1991;**179**:181-186. DOI: 10.1016/0009-2614(91) 90312-W

[56] Gallagher SH, Bolskar RD, Lay PA, Reed CA. Raman excitation profiles of C_{70} in benzene solutions. Assignment of electronic spectrum in the 380-510 nm region. Journal of the American Chemical Society. 1997;**119**:4263-4271. DOI: 10.1021/ja963426f

[57] Sun G, Kertesz M. Vibrational Raman spectra of C_{70} and studied by density functional theory. Journal of Physical Chemistry A. 2002;**106**:6381-6389. DOI: 10.1021/jp020222e

Chapter 6

Super-Lubricious, Fullerene-like, Hydrogenated Carbon Films

Bin Zhang, Kaixiong Gao, Yuanlie Yu and
Junyan Zhang

Additional information is available at the end of the chapter

http://dx.doi.org/10.5772/intechopen.70412

Abstract

Almost one-third of minimal energy is consumed via friction and wear process. Thus, to save energy using advanced lubrication materials is one of the main routes that tribologists are focused on. Recently, superlubricity is the most prominent way to face energy problems. Designing promising mechanical systems with ultra-low friction performance and establishing superlubricity regime is imperative not only to the most greatly save energy but also to reduce hazardous waste emissions into our environment. At the macroscale, hydrogenated diamond-like carbon (DLC) film with a supersmooth and fully hydrogen terminated surface is the most promising materials to realize superlubricity. However, the exact superlubricity of DLC film can only be observed under high vacuum or specific conditions and is not realized under ambient conditions for engineering applications. The latest breakthrough in macroscale superlubricity is made by introducing fullerene-like nano-structure and designing graphene nanoscroll formation, which also demonstrates the structure-superlubricity (coefficient of friction ~0.002) relationship. Thus, it is very interesting to design macroscale superlubricity by prompting the in situ formation of these structures at the friction interfaces. In this chapter, we will focus on fullerene-like hydrogenated carbon (FL-C:H) films and cover the growth methods, nanostructures, mechanic, friction properties and superlubricity mechanism.

Keywords: fullerene-like, carbon films, superlubricity, friction

1. Introduction

People have a love and hate relationship with friction because we need high friction some times (like braking) but we expect the friction to be as small as possible in the machines which might help to save energy. The energy issue pushes us to develop more robust lubricious materials. Nowadays, superlubricity is a most fascinating word not only in mind but also in practice [1–5]. Designing promising mechanical systems with ultra-low friction performance and establishing superlubricity regime is imperative not only to the most greatly save energy but also to reduce hazardous waste emissions.

Motivated by these issues, superlubricity has aroused extensive attention of many research groups and therefore an active topic in many fields [6–11]. Not surprisingly, superlubricity has been realized in some experiments associated with layered materials such as graphene, molybdenum disulphide (MoS_2), highly oriented pyrolytic graphite (HOPG) and multi-walled carbon nanotubes (MWCNT) [6, 12–15]. At the macroscale, hydrogenated diamond-like carbon (DLC) film with a supersmooth and fully hydrogen-terminated surface is the most promising material to realize superlubricity [4, 16]. However, the exact superlubricity of DLC film can only be observed under high vacuum or specific conditions and is not realized under ambient conditions for engineering applications [17]. The latest breakthrough in macroscale superlubricity is made by introducing fullerene-like nano-structure and designing graphene nanoscroll formation, which also demonstrates the structure-superlubricity (coefficient of friction ~0.002) relationship [5, 18].

In general, lubricating materials satisfying engineering application face some problems such as macro contact area (≥ mm × mm), withstanding high contact pressure (≥1 GPa) and exposure to air environment (H_2O, O_2, etc.), which are more prominent problems for designing and realizing macroscale superlubricity [19–24]. But, luckily, Zhang et al. fabricated fullerene-like hydrogenated carbon (FL-C:H) films and achieved superlubricity under engineering conditions [5, 18, 24–28]. The curvature structure of fullerene-like structure extends the strength of graphite plane hexagon into three dimension space network, in turn, increasing the hardness and elasticity of carbon films, demonstrating super low friction in air, meaning solid superlubricity with engineering application value [18, 26]. Most interestingly, the fullerene-like structure of the carbon films could be adjusted via the hydrogen content, bias supply, in the deposition gas sources [29–33]. Thus, it is very interesting to design macroscale superlubricity by prompting the in situ formation of these structures at the friction interfaces.

Usually, both plasma-enhanced chemical vapor deposition (PECVD) and reactive magnetron sputtering can be employed to growth FL-C:H films which one can find in early reports [5, 27]. When using PECVD, the pulse power and the reasonable atmosphere are necessary. FL-C:H films are growing in miexed atmosphere of CH_4 and H_2 with flow ratio at 1:2, but the deposition pressure can adjust from 10 to 20 Pa depending on different chambers [5, 34]. But for reactive magnetron sputtering, distinction from PECVD are that the additional magnetic field and no H_2 anymore [21, 27]. The most important thing when using reactive magnetron sputtering is that a deep poisoning mode is employed [27, 28]. So, as we see, deep poisoning reactive magnetron sputtering can also be mentioned as magnetic field assistant PECVD. Such difference might influence the growth mechanisms which we will discuss later.

2. Structural characterizations

As one known, carbon-based curve structure is variety of things, such as carbon nanotubes, fullerenes, carbon nitride nanotubes and so on [35, 36]. FL-C:H films and its structures are very similar to fullerene-like carbon nitride (FL-CNx) films [37], within distorted multi-storey multilayer graphene intersecting and interlocked in an amorphous carbon matrix, show high hardness, high elastic recovery and high resistance to deformation. A typical high resolution transmission electron images (HRTEM) of FL-C:H films, grown via high impulse power (which is usually employed in magnetron sputtering, mentioned as high power impulse magnetron sputtering (HiPIMS) [26]) assistant PECVD, displays in **Figure 1a**. Atomic force acoustic microscope (AFM) imaging (**Figure 1b**) under tapping mode was used to inspect the cluster domains in the films and indicate that the film has a special nanocomposite structure with graphene domains dispersed in an amorphous carbon matrix [26].

These special nanostructures can also be inspected by Raman spectrum. Usually, FL-C:H films exhibit typical character of diamond-like carbon films in the region of 1000–2000 cm^{-1}, but with some additional peaks at about 400 and 700 cm^{-1} and a distinct shoulder at around 1230 cm^{-1} **(Figure 2a)** [34]. The two low intermediate wave number bands near 400 cm^{-1} and 700 cm^{-1} are very similar to that of fullerene-like carbon nitride that can be assigned to relaxation of Raman selection rule due to the curvature in graphene planes, which appears to induce Raman scattering away from the G point [38]. The same bands are also present in the Raman spectra acquired from carbon onions and C_{60} that have been attributed to the transverse optic and transverse acoustic vibrations at the M point [38]. Thus, an acceptable fitting could only be reached via four Gaussian peaks at about 1230, 1350, 1470, and 1560 cm^{-1}, respectively **(Figure 2b)** [9, 27, 28, 34]. Thus, we can consider that all peaks at around 400, 700, 1230 and 1480 cm^{-1} are all active from that of fullerene-like structure with curled graphene planes.

Thanks to the adjustable conditions and methods, the nanostructures of FL-C:H films can be tailored as one expecting. In fact, either a conventional PECVD or a magnetic field assistant

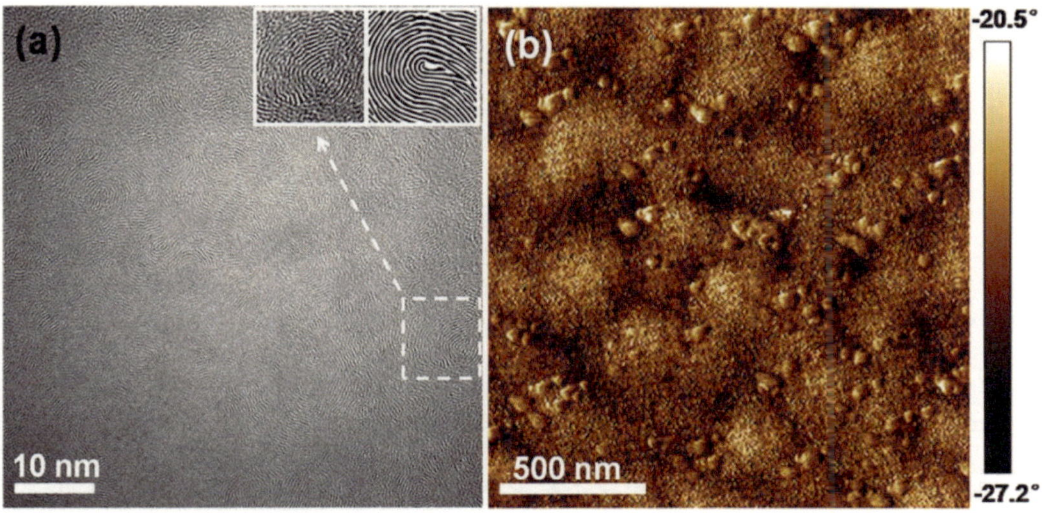

Figure 1. (a) High resolution transmission electron images of the FL-C:H films. Insets: carbon fingerprint and human fingerprint pattern. (b) An AFM phase image of the FL-C:H films deposited on silicon wafers. (Reproduced from Ref. [26] with permission from the Royal Society of Chemistry).

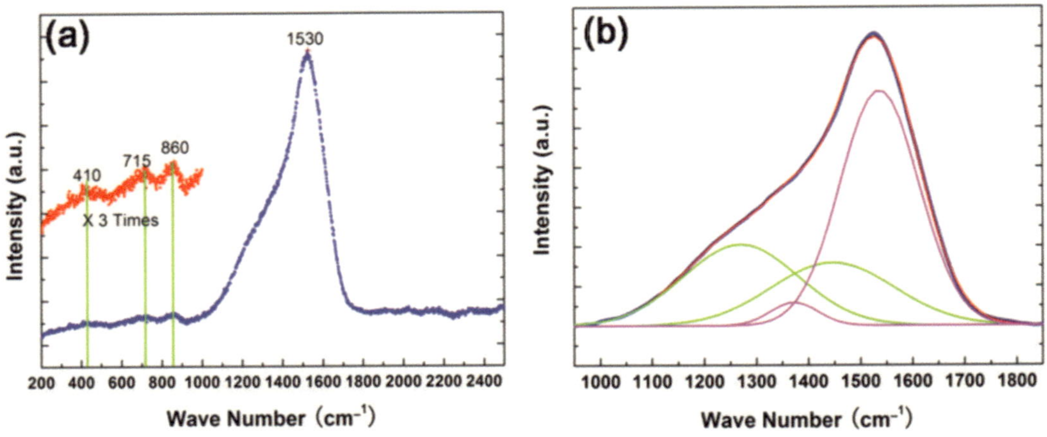

Figure 2. Raman spectra of hydrogenated carbon film deposited by dc-pulse plasma CVD. (a) Raman spectrum and magnified wave number region from 0 to 1000 cm^{-1} and (b) deconvoluted wave number region from 1000 to 2000 cm^{-1} in the Raman spectrum of (a).

PECVD (reactive magnetron sputtering), an important thing is that bias voltage has a main effects on nanostructures. Though the outfield auxiliary in reactive magnetron sputtering lowers the growth pressure directly, both high voltage and low duty cycle are crucial. During reactive magnetron sputtering process [21], the prominent FL-C:H films grown at −800 V bias exhibit the highest hardness of 20.9 GPa as well as elastic recovery of about 85%. The further increase of bias voltage cripples the mechanical due to further graphitization of the films. But very different nanostructure can be observed under low bias of −100 V, which is called amorphous carbon films dispersed with multilayer graphenes. It is very different from the films grown via PECVD that one can seen that with increasing the bias voltage, the hardness decreases while the elastic recovery keeps increasing [39]. The probable reason is that, at low pressure with the outfield auxiliary in reactive magnetron sputtering process, ions have higher free energy than that in PECVD which might induce easier graphitization.

On the other hand, the pulse duty cycles determine the local relaxation of the distorted chemical bonds. According to the typical subplantation model suggested by Robertson [40], three scales during ion interaction with the film are addressed that: the cascade, 10^{-14} s; the thermal spike, 10^{-12} s; and the longer time relaxation, ~10^{-9} s. The longer time relaxation time benefits the stress releasing and hydrogen removing which contributes to restructuring of carbon matrix. In our work, the annealing time is almost ~10^{-5} s, far than ~10^{-9} s, thus, the depositing pulse film has a completely surface, thereby restricting the formation of pentagonal rings which are associated with the ion bombardment at the pulse-on/off state [41]. As shown in **Figure 3**, one can see that with the pulse-on time increases while pulse-off time decrease, more curved graphene structures are arisen. These variations are also confirmed by Raman spectra, showing in **Figure 4**. With decrease in duty cycles, a shoulder at around 1230 cm^{-1} comes more obviously, which is believed from that of fullerene-like structure with curled graphene planes, in accordance well with HRTEM results. Here, Raman spectra of these films are simulated using four vibrational bands at 1260, 1380, 1470, and 1570 cm^{-1} [9, 27, 29, 34], respectively (**Figure 4**), these with A-type symmetry (from five-, six-, and seven-membered

http://dx.doi.org/10.5772/intechopen.70412

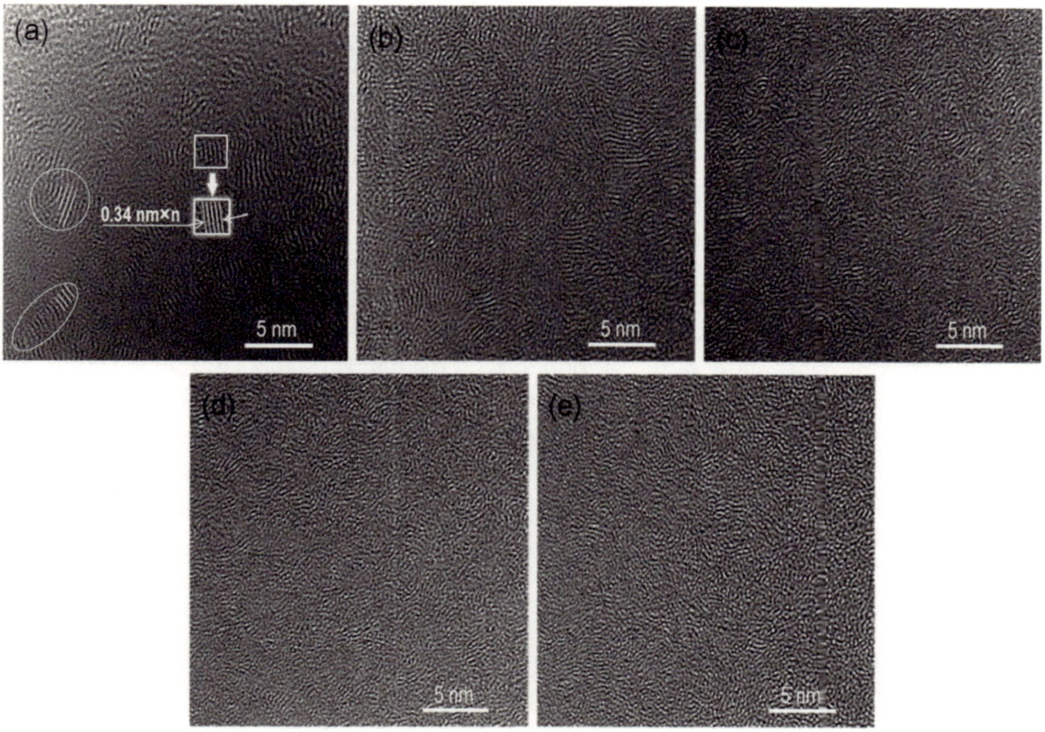

Figure 3. HRTEM images of the as-prepared films with pulse duty cycles of (a) 100%, (b) 80%, (c) 60%, (d) 40%, and (e) 20%. (Reproduced from Ref. [42] with permission from the Royal Society of Chemistry).

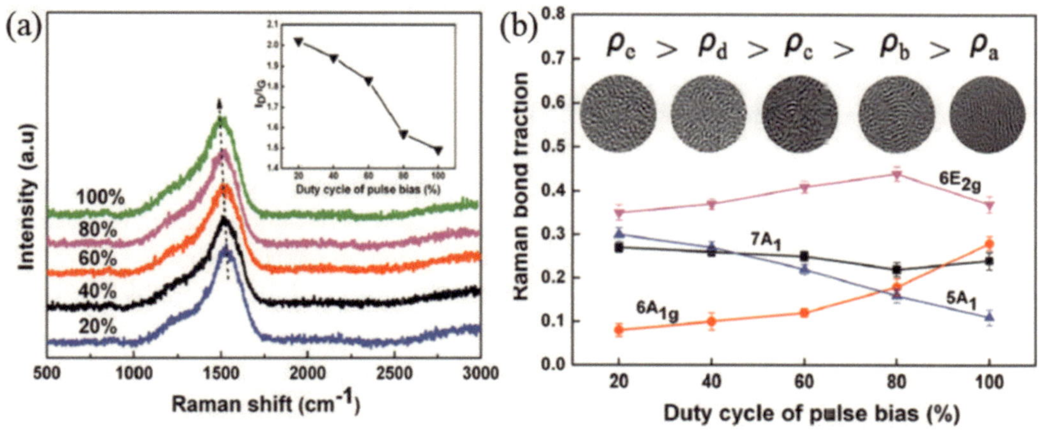

Figure 4. (a) Raman spectra of the as-prepared films at different pulse bias duty cycle. (b) Contribution to the carbon Raman band from the vibrations of five-, six- and seven-member rings versus the pulse duty cycle. (Reproduced from Ref. [42] with permission from the Royal Society of Chemistry).

rings) and one with E-type symmetry (from six-membered rings). The results show that even member rings and odd member rings have an opposite trend, and high odd member rings indicate the decrease and more fullerene-like structure.

Atmosphere influence on the nanostructures of FL-C:H films have been studied widely. Heterogeneous gases, such as H_2 and CF_4, introducing in growth process have different effects [29–33]. Interestingly, the growth of FL-C:H films show a conflicting to Hellgren's work [43] that intentional hydrogen addition to the discharge will terminate potential bonding sites for CNx precursors and hinder the growth of fullerene-like structures. But it seem that during growth FL-C:H films, hydrogen atoms during the deposition process may affect the production of odd rings by two competing ways: (1) stress induced by H^+ leads to the introduction of odd ring into flat graphene plane; (2) H^+ preferentially etches the plane's sp^2 phase and destroys the bond basis if forming odd rings. But more H_2 existing in growth condition has no much help on growing fullerene-like structures, so the effects of hydrogen need to be studied in detail. However, CF_4 show different influence on the nanostructures of the FL-C:H films. At low fluorine content, many C sites bond to neighboring C and the films microstructure displays lots of well organized graphite-like and fullerene-like fragments. But as the amount of F incorporated in the network increase, F-terminated large rings, Branches and chains with sp^2 sites densify and start to interact with each other and features like interlocking pore and amorphousness strongly prevail in the nanostructure [30, 32].

3. Mechanical and tribology properties

As we know, thanks to the unique topological structure, fullerene, predicted by Ruoff [44], have higher hardness than diamond, and have confirmed by other groups(is C60 fullerite harder than diamond). Vadim V. Brazhkin has synthesized a new nanomaterial by high-temperature treatment of fullerite C60 at moderate (0.1–1.5 GPa) pressures attainable for large-volume pressure apparatus, which show a high (90%) elastic recovery and fairly hardness of about 10–15 GPa [45]. It is noteworthy that in our works, so-called FL-C:H films with curved graphenes show a elastic recovery (≥80) with hardness variation from 10 to 30 GPa, depending on the growth conditions. **Figure 5** shows typical load-displacement curve and friction coefficients as function of time. A superlubricity phenomenal is observed at load of 20 N with friction coefficient at 0.009. One can notice the elastic recovery value variation with the annealing temperatures and shows the highest recovery value as shown in **Figure 6**. It is easily understood if one employs pentatomic-heptatomic theory of Raman [27, 28]. At 300°C, films show the maximum of odd rings which indicates that the most value of curved graphene exiting inside the carbon amorphous matrix.

As discussed above, it is obvious that elastic recovery and friction properties are correlated with inner nanostructures of FL-C:H films. To confirm this, we also studied the elastic recovery and friction properties using the films growth under different duty cycles and H_2 flow rates. As we have already confirmed that lower duty cycle means more fullerene-like structure due to the long relaxing time. So the lower duty cycle declares that lower growth rate of the films thickness (**Figure 7** inset). One can also confirm that the hardness and the elastic recovery decrease with the decay of fullerene-like structure, as the decrease of odd member rings (**Figure 6**).

http://dx.doi.org/10.5772/intechopen.70412

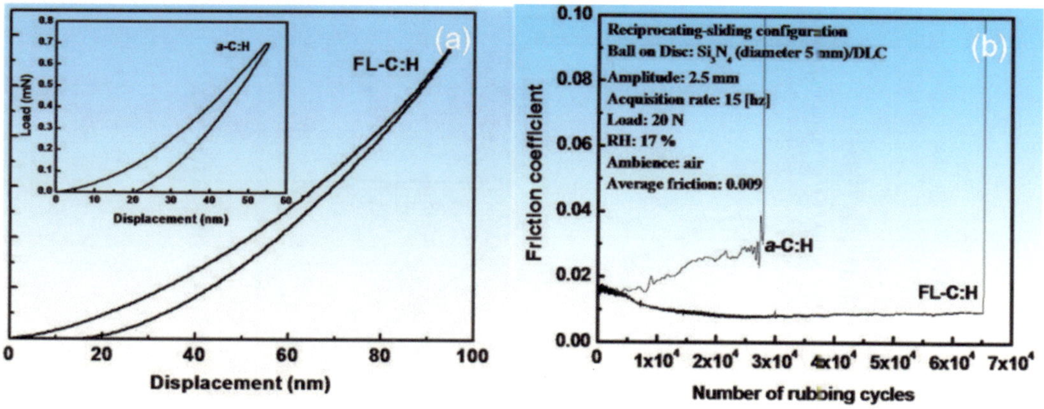

Figure 5. Typical load-displacement curve for the 1500 nm thick FL-C:H film annealed at 300°C, the inset is the load-displacement curve of 500 nm thick typical hydrogenated carbon films (a-C:H) (a) and tribological tests on the FL-C:H film in air with 20% relative humidity at ambient temperature (b).

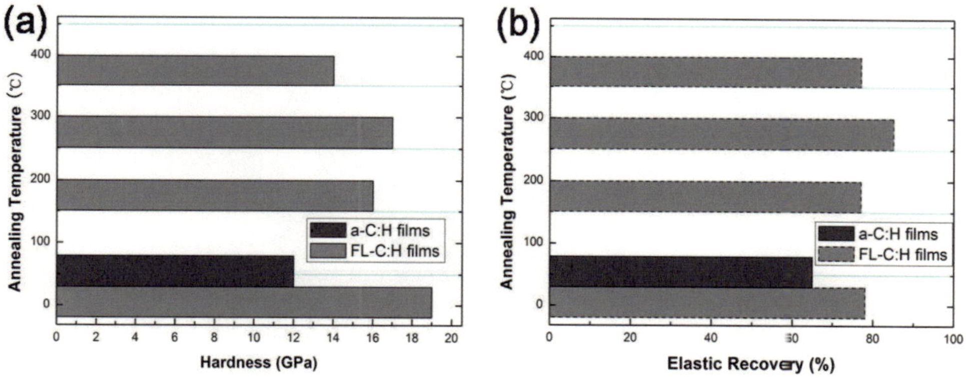

Figure 6. Mechanical properties of the FL-C:H film before and after annealing: (a) Hardness, and (b) Elastic Recovery. (All films were annealed for 1h in Ar at temperature of 200, 300 and 400°C, respectively).

Different from the single-way variation of the influence of duty cycle, the change of mechanical properties has a yielding point. As shown in **Figure 8**, the FL-C:H film grown at H_2 flow rate 5 SCCM show the lowest friction coefficient. Raman spectrum is employed here to quantify the odd member rings from both films and corresponding debris. It is showing that the friction coefficient has a significantly positive relation with odd member rings. That is, odd member rings confirm more fullerene-like structures which in turn, determine the hardness and tribology. So there is no excuse that fullerene-like structures contribute to mechanical and tribology properties.

To reveal how such structure affects on the tribology properties, X-ray diffraction (XRD) analysis has been performed at the original surface and wear debris and tracks of the FL-C:H films. The film pattern shows three peaks at about $2\theta = 69.1°$, 33° and 22.4°. The two peaks at $2\theta = 69.1°$ and 33° are from the silicon substrate ([004] and [002], respectively (**Figure 9**)). A

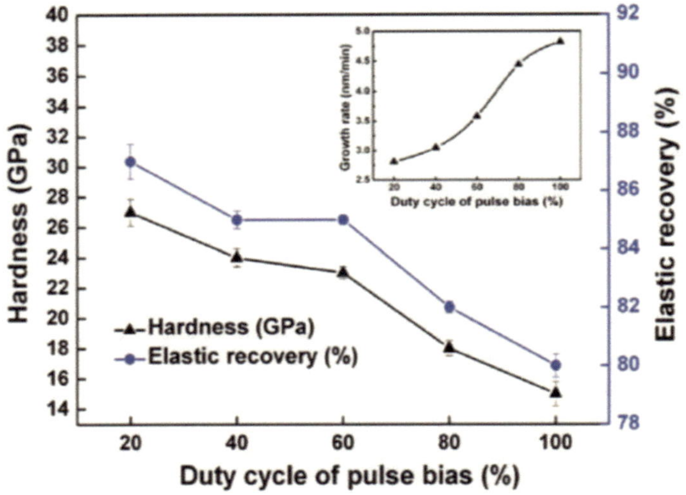

Figure 7. Hardness and elastic recovery of the as-prepared films with different pulse duty cycles. Inset show the growth rate of the as-prepared films with different pulse duty cycles. (Reproduced from Ref. [42] with permission from the Royal Society of Chemistry).

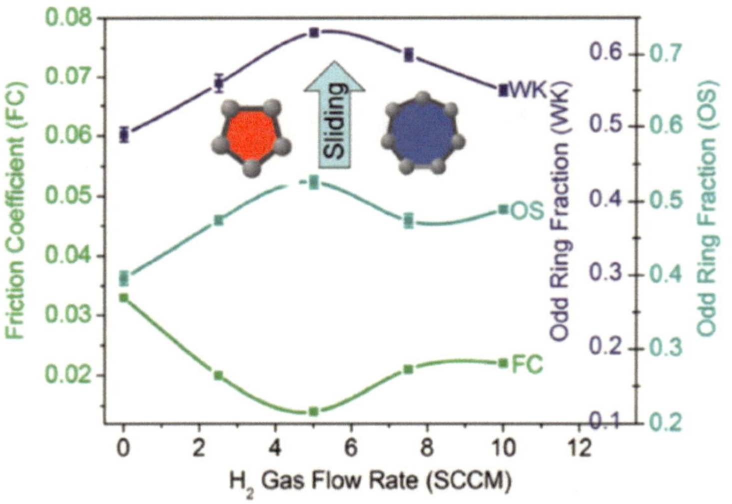

Figure 8. Friction coefficient and odd ring fraction of the FL-C:H films as a function of the gas flow rate of H_2. All the central wear tracks (WK) have a higher odd (pentagonal and heptagonal) carbon ring fraction than that of the originally deposited surfaces (OS). (Reproduced from Ref. [45] with permission from the Royal Society of Chemistry).

weak peak at $2\theta = 22.4°$, according with other studies [46], can be attributed to fullerene-like or onion-like nanoparticles (considering HRTEM and Raman results). And after friction testing, the peak at $2\theta = 22.4°$ become prominent, accompanying with a new band at $2\theta = 15°$ which arise from fullerene-like or onion-like nanoparticles, which leads low friction and small wear in open wear.

http://dx.doi.org/10.5772/intechopen.70412

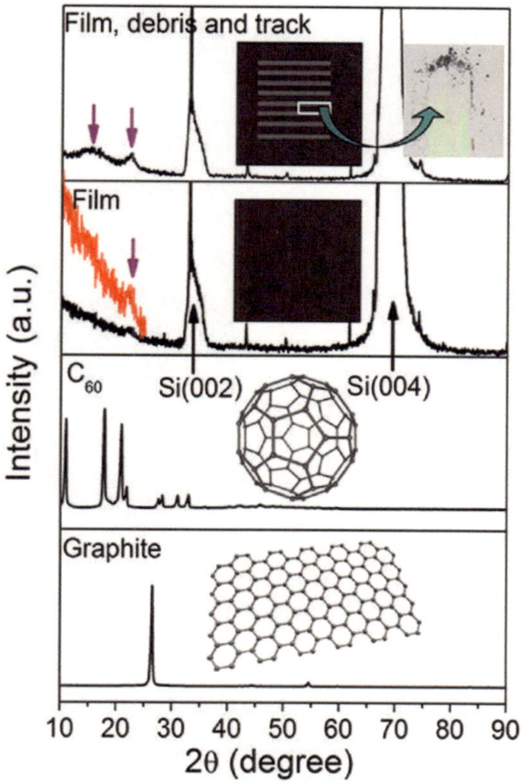

Figure 9. Experiment XRD patterns from the film and wear debris showing that the debris has a structure analogous to that of C60, far from that of nc-graphite. (Reproduced from Ref. [45] with permission from the Royal Society of Chemistry).

Besides, the affects of F incorporation, the humidity, variation of load, plasma process, etc., on the tribology properties of FL-C:H films were widely studied [21, 23–27]. Unfortunately, though the introducing of F atoms in carbon matrix is active hydrophobicity properties, and destroys the fullerene-like structures via terminating [30–32]. But humidity has a great influence on friction coefficients that, with increasing the humidity, the friction coefficient increases quickly to 0.08 at 50% for humidity. HRTEM results show that the onion-like nanoparticles in the debris are restrained: from spherical shell structure below 30% humidity to short curved graphene dispersed in amorphous matrix [26].

4. Superlubricity mechanisms

Benefit from unique bulk structures of FL-C:H films, some of well shelled graphene assemble nanoparticles are discovered on the friction interface. Thus, one can speculate that the main reason to the superlubricity is based on the bulk of fullerene-like structure which provides the

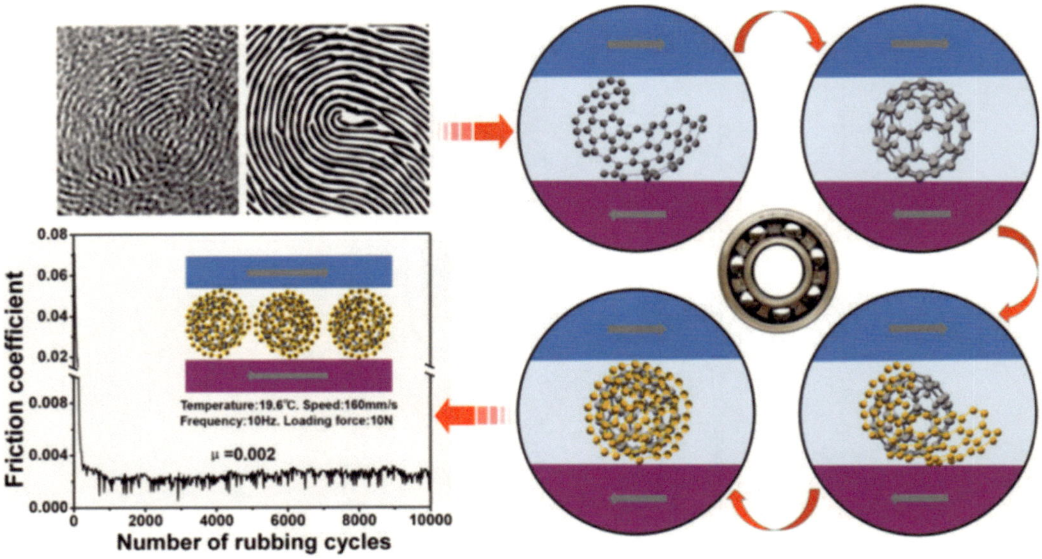

Figure 10. A incommensurate and rolling contact for the superlubricity of FL-C:H films. (Reproduced from Ref. [26] with permission from the Royal Society of Chemistry).

raw materials to grow of onion-like nanoparticles at certain conditions. One of issues worth exploring is that the onion-like nanoparticles only can see below 30% humidity. Theoretical model of the graphene to fullerene transformation confirmed that the formation of defects at the edge of graphene is the crucial step. Usually, water molecules exhibit higher activation energies that those of the non-polar adsorbates (O_2, N_2 and Ar) via stronger hydrogen bonds. Thus, at lower humidity, unsaturated dangling bonds of graphene have a high reaction activity and the hybridization are easy to occur on the edge of graphene itself, but at higher humidity, dangling bonds are saturated by water molecules, hindering the hybridization of graphene fragment, which limits the nucleation and growth of onion-like nanoparticles. In addition, these particles have a chemically inert surface, which reduces the couple of dual interface via offering incommensurate and rolling contact (**Figure 10**).

5. Conclusions

FL-C:H films have been grown via PECVD and reactive magnetron sputtering methods. Such films with curved graphene dispersed in amorphous matrix, show elastic recovery (≥80), hardness (variation from 10 to 30 GPa) as well as low friction properties. Sometimes, FL-C:H films depict superlubricity within limited situation with the lowest coefficient value to 0.002. We believe that superlubricity of FL-C:H films benefits from the unique fullerene-like structure, which provide raw materials to grow onion-like nanoparticles during friction tests. In addition, these particles have a chemically inert surface, which reduces the couple of dual interface via offering incommensurate and rolling contact. But, further developing of the superlubricity properties of the FL-C:H films to wide adaptability is very important to broad practicability.

Acknowledgements

This work is supported by CAS "Light of West China" Program, Youth Innovation Promotion Association CAS (Grant no. 2017459), China Scholarship Council (File no. 201604910183) and the National Natural Science Foundation of China (Grant nos. 51205383, 51611530704 and 51661135022).

Author details

Bin Zhang[1,2,3]*, Kaixiong Gao[1,2], Yuanlie Yu[1,2] and Junyan Zhang[1,2]

*Address all correspondence to: bzhang@licp.cas.cn

1 R&D Center of Lubricating and Protecting Materials, Lanzhou Institute of Chemical Physics, Chinese Academy of Science, Lanzhou, Gansu, P.R. China

2 Key Laboratory of Solid Lubrication, Lanzhou Institute of Chemical Physics, Chinese Academy of Science, Lanzhou, Gansu, P.R. China

3 Lawrence Berkeley National Laboratory, Berkeley, CA, USA

References

[1] Hirano M, Shinjo K. Atomistic locking and friction. Physical Review B. 1990;**41**(17):11837. DOI: 10.1103/PhysRevB.41.11837

[2] Li J, Luo J. Advancements in superlubricity. Science China Technological Sciences. 2013;**56**(12):2877-2887. DOI: 10.1007/s11431-013-5387-y

[3] Berman D, Deshmukh SA, Sankaranarayanan SKRS, Erdemir A, Sumant AV. Macroscale superlubricity enabled by graphene nanoscroll formation. Science. 2015;**348**(6239):1118-1122. DOI: 10.1126/science.1262024

[4] Erdemir A. Design criteria for superlubricity in carbon films and related microstructures. Tribology International. 2004;**37**(7):577-583. DOI: 10.1016/j.triboint.2003.12.007

[5] Wang C, Yang S, Wang Q, Wang Z, Zhang J. Super-low friction and super-elastic hydrogenated carbon films originated from a unique fullerene-like nanostructure. Nanotechnology. 2008;**29**(22):225709. DOI: 10.1088/0957-4484/19/22/225709

[6] Martin JM, Pascal H, Donnet C, Le Mogne T, Loubet JL, Epicier T. Superlubricity of MoS2: Crystal orientation mechanisms. Surface and Coatings Technology. 1994;**68-69**: 427-432. DOI: 10.1016/0257-8972(94)90197-X

[7] Lantz MA, Wiesmann D, Gotsmann B. Dynamic superlubricity and the elimination of wear on the nanoscale. Nature Nanotechnology. 2009;**4**:586-591. DOI: 10.1038/nnano.2009.199

[8] Li H, Wood RJ, Rutland MW, Atkin R. An ionic liquid lubricant enables superlubricity to be "switched on" in situ using an electrical potential. Chemical Communications. 2014;**50**:4368-4370. DOI: 10.1039/C4CC00979G

[9] Li J, Zhang C, Ma L, Liu Y, Luo J. Superlubricity achieved with mixtures of acids and glycerol. Langmuir. 2013;**29**(1):271-275. DOI: 10.1021/la3046115

[10] Itamura N, Miura K, Sasak N. Simulation of scan-directional dependence of superlubricity of C60 molecular bearings and graphite. Japanese Journal of Applied Physics. 2009;**48**:060207. DOI: 10.1347-4065/48/6R/060207

[11] Hirano M, Shinjo K, Kaneko R, Murata Y. Observation of superlubricity by scanning tunneling microscopy. Physical Review Letters. 1997;**78**:1448. DOI: 10.1103/PhysRevLett.78.1448

[12] Donnet C, Mogne TL, Martin JM. Superlow friction of oxygen-free MoS2 coatings in ultrahigh vacuum. Surface and Coatings Technology. 1993;**62**(1-3):406-411. DOI: 10.1016/0257-8972(93)90275-S

[13] Dienwiebel M, Verhoeven GS, Pradeep N, Frenken JWM, Heimberg JA, Zandbergen HW. Superlubricity of graphite. Physical Review Letters. 2004;**92**:126101. DOI: 10.1103/PhysRevLett.92.126101

[14] Liu Z, Yang J, Grey F, Liu JZ, Liu Y, Wang Y, Yang Y, Cheng Y, Zheng Q. Observation of microscale superlubricity in graphite. Physical Review Letters. 2012;**108**:205503. DOI: 10.1103/PhysRevLett.108.205503

[15] Zhang R, Ning Z, Zhang Y, Zheng Q, Chen Q, Xie H, Zhang Q, Qian W, Wei F. Superlubricity in centimetres-long double-walled carbon nanotubes under ambient conditions. Nature Nanotechnology. 2013;**8**:912-916. DOI: 10.1038/nnano.2013.217

[16] Erdemir A, Eryilmaz OL, Fenske G. Synthesis of diamond like carbon films with superlow friction and wear properties. Journal of Vacuum Science & Technology A: Vacuum, Surfaces, and Films. 2000;**18**(4):582459. DOI: 10.1116/1.582459

[17] Andersson J, Erck RA, Erdemi A. Friction of diamond-like carbon films in different atmosphere. Wear. 2002;**51**:1070-1075. DOI: 10.1016/S0043-1648(03)00336-3

[18] Gong Z, Shi J, Zhang B, Zhang J. Graphene nano scrolls responding to superlow friction of amorphous carbon. Carbon. 2017;**116**:310-317. DOI: 10.1016/j.carbon.2017.01.106

[19] Erdemir A, Donnet C. Tribology of diamond-like carbon films: Recent progress and future prospects. Journal of Physics D: Applied Physics. 2006;**39**(18):R311-R327. DOI: 10.1088/0022-3727/39/18/R01

[20] Andersson J, Erck RA, Erdemir A. Frictional behavior of diamond like carbon films in vacuum and under varying water vapor pressure. Surface and Coatings Technology. 2003;**163**:535-540. DOI: 10.1016/S0257-8972(02)00617-5

[21] Wang X, Wang P, Zhang B, Yang S, Zhang J. The tribological properties of fullerene-like hydrogenated carbon (FL-C: H) film under different humidity conditions. Tribology Transactions. 2009;**52**(3):354-359. DOI: 10.1080/10402000802563125

[22] Zhou S, Wang L, Xue Q. Achieving low tribological moisture sensitivity by aC: Si: Al carbon-based coating. Tribology Letters. 2001;**43**:329-339. DOI: 10.1007/s11249-011-9814-6

[23] Shi J, Gong Z, Wang C, Zhang B, Zhang J. Tribological properties of hydrogenated amorphous carbon films in different atmospheres. Diamond and Related Materials. 2017;**77**:84-91. DOI: 10.1016/j.diamond.2017.06.005

[24] Wang Y, Zhang B, Gong Z, Gao K, Ou Y, Zhang J. The effect of a static magnetic field on the hydrogen bonding in water using frictional experiments. Journal of Molecular Structure. 2013;**1052**:102-104. DOI: 10.1016/j.molstruc.2013.08.021

[25] Gong Z, Jia X, Ma W, Zhang B, Zhang J. Hierarchical structure graphitic-like/MoS_2 film as superlubricity material. Applied Surface Science. 2017;**413**:381-386. DOI: 10.1016/j.apsusc.2017.04.057

[26] Gong Z, Shi J, Ma W, Zhang B, Zhang J. Engineering-scale superlubricity of the fingerprint-like carbon films based on high power pulsed plasma enhanced chemical vapor deposition. RSC Advances. 2016;**6**(116):115092-115100. DOI: 10.1039/C6RA24933G

[27] Wang P, Wang X, Zhang B, Liu W. Structural, mechanical and tribological behavior of fullerene-like carbon film. Thin Solid Films. 2010;**518**(21):5938-5943. DOI: 10.1016/j.tsf.2010.05.096

[28] Wang P, Wang X, Zhang B, Liu W. Formation of hydrogenated amorphous carbon films containing fullerene-like structures. Journal of Non-Crystalline Solids. 2009;**355**(34):1742-1746. DOI: 10.1016/j.jnoncrysol.2009.06.014

[29] Wang Y, Guo J, Gao K, Zhang B, Liang A, Zhang J. Understanding the ultra-low friction behavior of hydrogenated fullerene-like carbon films grown with different flow rates of hydrogen gas. Carbon. 2014;**77**:518-524. DOI: 10.1103/PhysRevB.75.205402

[30] Zhang L, Wang J, Zhang J, Zhang B. Increasing fluorine concentration to control the microstructure from fullerene-like to amorphous in carbon films. RSC Advances. 2016;**6**(26):21719-21724. DOI: 10.1039/C6RA00675B

[31] Qiang L, Zhang B, Gao K, Gong Z, Zhang J. Hydrophobic, mechanical, and tribological properties of fluorine incorporated hydrogenated fullerene-like carbon films. Friction. 2013;**1**(4):350-358. DOI: 10.1007/s40544-013-0031-1

[32] Wei L, Zhang B, Zhou Y, Qiang L, Zhang J. Ultra-low friction of fluorine-doped hydrogenated carbon film with curved graphitic structure. Surface and Interface Analysis. 2013;**45**(8):1233-1237. DOI: 10.1002/sia.5260

[33] Liu X, Yang J, Hao J, Zheng J, Gong Q, Liu W. A near-frictionless and extremely elastic hydrogenated amorphous carbon film with self-assembled dual nanostructure. Advanced Materials. 2012;**24**:4614-4617. DOI: 10.1002/adma.201200085

[34] Wang Q, Wang C, Wang Z, Zhang J. Fullerene nanostructure-induced excellent mechanical properties in hydrogenated amorphous carbon. Applied Physics Letters. 2007;**91**(14):141902. DOI: 10.1063/1.2794017

[35] Yan L, Zhao F, Li S, Hu Z, Zhao Y. Low-toxic and safe nanomaterials by surface-chemical design, carbon nanotubes, fullerenes, metallofullerenes, and graphenes. Nanoscale. 2011;**3**:362-382. DOI: 10.1039/C0NR00647E

[36] Zhang J, Zhang B, Xue Q, Wang Z. Ultra-elastic recovery and low friction of amorphous carbon films produced by a dispersion of multilayer graphene. Surface and Interface Analysis. 2012;**44**(2):162-165. DOI: 10.1002/sia.3787

[37] Liu DG, Tu JP, Gu CD, Hong CF, Chen R, Yang WS. Synthesis, structure and mechanical properties of fullerene-like carbon nitride films deposited by DC magnetron sputtering. Surface and Coatings Technology. 2010;**205**(7):2474-2482. DOI: 10.1016/j.surfcoat.2010.09.043

[38] Roy D, Chhowalla M, Wang H, Sano N, Alexandrou I, Clyne TW, Amaratunga GAJ. Characterisation of carbon nano-onions using Raman spectroscopy. Chemical Physics Letters. 2003;**373**(1-2):52-56. DOI: 10.1016/S0009-2614(03)00523-2

[39] Wang J, Cao Z, Pan F, Wang F, Liang A, Zhang J. Tuning of the microstructure, mechanical and tribological properties of a-C:H films by bias voltage of high frequency unipolar pulse. Applied Surface Science. 2015;**356**(30):695-700. DOI: 10.1016/j.apsusc.2015.08.091

[40] Robertson J. Mechanism of sp3 bond formation in the growth of diamond-like carbon. Diamond and Related Materials. 2005;**14**(3-7):942-948. DOI: 10.1016/j.diamond.2004.11.028

[41] Wang Y, Gao K, Shi J, Zhang J. Bond topography and nanostructure of hydrogenated fullerene-like carbon films: A comparative study. Chemical Physics Letters. 2016;**660**(1):160-163. DOI: 10.1016/j.cplett.2016.08.023

[42] Liu G, Zhou Y, Zhang B, Gao K, Qiang L, Zhang J. Monitoring the nanostructure of a hydrogenated fullerene-like film by pulse bias duty cycle. RSC Advances. 2016;**6**(64):59039-59044. DOI: 10.1016/j.cplett.2016.08.023

[43] Hellgren N, Johansson MP, Hjörvarsson B, Östblom M, Liedberg B, Hultman L, Sundgren J-E. Growth, structure, and mechanical properties of CNxHyCNxHy films deposited by dc magnetron sputtering in N2/Ar/H2 discharges. Journal of Vacuum Science & Technology A: Vacuum, Surfaces, and Films. 2000;**18**:2349. DOI: 10.1116/1.1286395

[44] Ruoff RS. Is C60 stiffer than diamond? Nature. 1991;**350**(6320):663-664. DOI: 10.1038/350663b0

[45] Brazhkin VV, Solozhenko VL, Bugakov VI, Dub SN, Kurakevych OO, Kondrin MV, Lyapin AG. Bulk nanostructured carbon phases prepared from C60: Approaching the 'ideal' hardness. Journal of Physics: Condensed Matter. 2007;**19**(23):236209. DOI: 10.1088/0953-8984/19/23/236209

[46] Wang Y, Guo J, Zhang J, Qin Y. Ultralow friction regime from the in situ production of a richer fullerene-like nanostructured carbon in sliding contact. RSC Advances. 2015;**5**:106476-106484. DOI: 10.1039/C5RA20892K

Chapter 7

How Important is Metal-Carbon Back-Bonding for the Stability of Fullerene-Transition Metal Complexes? Role of Cage Sizes, Encapsulated Ions and Metal Ligands

Ming-Chung Yang and Ming-Der Su

Additional information is available at the end of the chapter

http://dx.doi.org/10.5772/intechopen.70068

Abstract

A density functional study of $\{\eta^2\text{-}(X@C_n)\}ML_2$ complexes with various cage sizes (C_{60}, C_{70}, C_{76}, C_{84}, C_{90}, C_{96}), encapsulated ions ($X = F^-$, 0, Li^+) and metal fragments (M = Pt, Pd) is performed, using M06/LANL2DZ levels of theory. The importance of π back-bonding to the thermodynamic stability of fullerene-transition metal complexes ($\{\eta^2\text{-}(X@C_n)\}ML_2$) and the effect of encapsulated ions, metal fragments and cage sizes on the π back-bonding are determined in this study. The theoretical computations suggest that π back-bonding plays an essential role in the formation of fullerene-transition metal complexes. The theoretical evidence also suggests that there is no linear correlation between cage sizes and π back-bonding, but the encapsulated Li^+ ion enhances π back-bonding and F^- ion results in its deterioration. These computations also show that a platinum center produces stronger π back-bonding than a palladium center. It is hoped that the conclusions that are provided by this study can be used in the design, synthesis and growth of novel fullerene-transition complexes.

Keywords: fullerene-transition metal complex, π back-bonding, encapsulated ions, metal fragments and cage sizes

1. Introduction

The first fullerene-transition metal complex, $(\eta^2\text{-}C_{60})Pt(Ph_3)_2$, was prepared and structurally characterized by Fagan et al. in 1991 [1]. It was the starting point for a new class of study for fullerene chemistry. Since then, various fullerene-transition metal complexes have been synthesized and these have potential applications in solar cells, spintronics, catalysis and drug delivery [2]. Balch et al. then studied the reactions of C_{60} with electron-rich fragments, $IrCl(CO)(PPh_3)_2$ and produced the fullerene-iridium complex $(\eta^2\text{-}C_{60})IrCl(CO)(PPh_3)_2$ [3]. The formation of fullerene-iridium complex is a reversible process and the reversible binding of $IrCl(CO)(PPh_3)_2$ to fullerenes can be used as a structural probe because the adducts can build ordered single crystals that are suitable for X-ray diffraction [4, 5]. Fullerene-iridium complexes that contain an enantiomeric phosphine ligand are used as solar photoelements [6]. One of the significant characteristics of fullerenes is that they are capable of encaging atoms, ions and small molecules to form endohedral complexes. Endohedral metallofullerenes (EMFs) are those that encapsulate metal atoms within a hollow carbon cage. Proft et al. theoretically studied the interactions between encapsulated monoatomic ions (Li^+ to Rb^+ and F^- to I^-) and C_{60} and its Si and Ge analogues [7] and found that, for these families, the interactions between $Li^+(Na^+)$ and $F^-(Cl^-)$ ions and C_{60} are strongest and exothermic, which confirms the possibility of the existence of these species. Recently, $Li^+@C_{60}$ was successfully synthesized and isolated by Watanabe et al. [8].

Understanding the strength and nature of metal-ligand bonding is crucial for the design of new fullerene-transition metal complexes because the structure and stability of various intermediates are important to the formation of organometallics [9]. In an earlier work by the authors [10], $\{\eta^2\text{-}(X@C_{60})\}ML_2$ complexes (M = Pt, Pd; X = 0, Li^+, L = PPh_3) were studied and it was found that there is a relationship between thermodynamic stability and π back-bonding; that is, the greater the π back-bonding, the greater is thermodynamic stability. This shows that thermodynamic stability can be modified by tuning the π back-bonding. As far as the authors are aware, π back-bonding could be affected by several factors, including the encapsulated ions, the metal fragments and the cage sizes. This study determines the importance of π back-bonding to the thermodynamic stability of $\{\eta^2\text{-}(X@C_n)\}ML_2$ complexes by using M = Pt, Pd; X = F^-, 0, Li^+ and n = 60, 70, 76, 84, 90 and 96 to ascertain the role of these factors in π back-bonding. Since the system is very large, methyl-substituted N-heterocyclic carbenes (NHC) are used as a ligand (L), instead of PPh_3. NHC is one of the frequently used and most powerful tools in organic chemistry [11]. In this work, the following reactions are studied:

$$ML_2 + X@C_n \rightarrow \{\eta^2\text{-}(X@C_n)\}\, ML_2 \qquad (1)$$

2. Computational details

The following fullerenes that comply with the isolated pentagon rule are used to develop a correlation: Ih-C_{60}, D_{5h}-C_{70}, D_2-C_{76}, D_{2d}(23)-C_{84}, D_{5h}(1)-C_{90} and D_{3d}(3)-C_{96}. These are experimentally isolated and identified [12–14]. The symmetry and numbering scheme for fullerene isomers are in accordance with an approved classification [15]. Hückel molecular orbital calculations show that the 6:6 ring junctions at the poles of the molecules usually have highest π bond orders (B) and are expected to be the most reactive, so these are the sites of attack (see **Scheme 1**) [12, 16].

http://dx.doi.org/10.5772/intechopen.70068

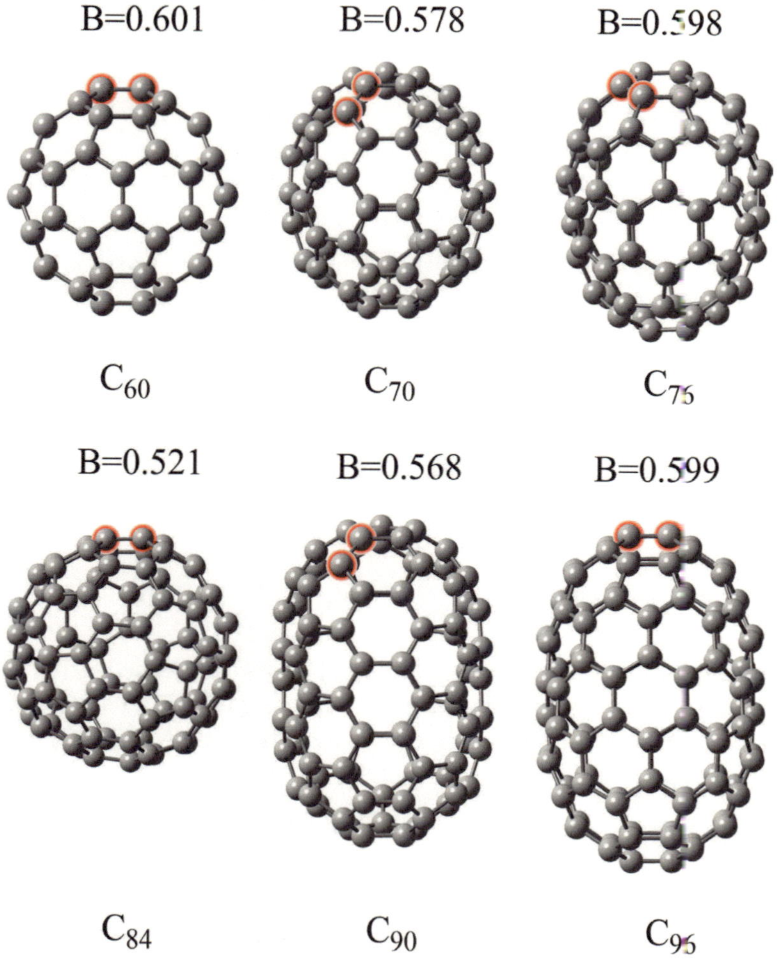

Scheme 1. The sites of attack for addition to the fullerenes Ih-C_{60}, D_{5h}-C_{70}, D_2-C_{76}, D_{2d}(23)-C_{84}, D_{5h}(1)-C_{90} and D_{5h}(1)-C_{96}. The Hückel π bond orders (B) were calculated using the freeware program, HuLiS [16].

The geometry optimizations are performed without any symmetry restrictions by using the M06 [17]/LANL2DZ [18, 19] level of theory. The vibrational frequency calculations at 298.15 K and 1 atm use the same level of theory. The stationary points are confirmed by the absence of imaginary frequencies. The natural charges are obtained using NBO 5.9, as implemented in the Gaussian 09 program [20].

The interatomic interactions are determined using energy decomposition analysis (EDA). Two types of EDA are used in this work. The first is the basic EDA that was developed individually by Yang et al. [21] and by Ziegler and Rauk [22]. For this basic EDA, the bonding energy (ΔE) is partitioned into two terms, ΔE = ΔE(DEF) + ΔE(INT). In this work, basic EDA is used for the optimized $ML_2X@C_n$ complexes, which are categorized into transition metal complexes (A), carbon cages (B) and metal ions (C) as shown in **Scheme 2**. The deformation energy (ΔE(DEF))

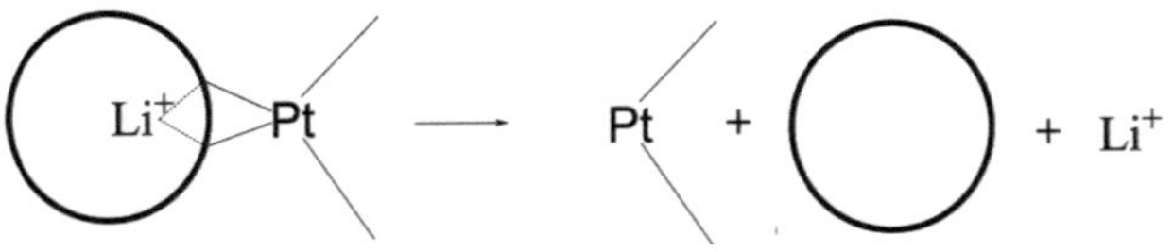

(ABC) A B C

$$\mathrm{Ao+Bo+C} \longrightarrow \underset{\Delta E(DEF)}{\mathrm{A+B+C}} \longrightarrow \mathrm{A+(BC)} \longrightarrow \underset{\Delta E(INT)_{A(BC)}}{\mathrm{(ABC)}}$$

$$\begin{aligned}\Delta E(DEF) &= (E_A+E_B+E_C)-(E_{Ao}+E_{Bo}+E_C)\\ &= (E_A-E_{Ao})+(E_B-E_{Bo})\\ &= \Delta E(DEF)_A+\Delta E(DEF)_B\end{aligned}$$

$$\Delta E(INT)_{A(BC)} = (E_{(ABC)})-(E_A+E_{(BC)})$$

Scheme 2. Basic EDA for $\{\eta^2\text{-}(Li^+@C_n)\}PtL_2$.

is the sum of the deformation energy of A ($\Delta E(DEF)_A$, which is defined as the energy of A in the product relative to the optimized isolated structure (A_0) and B ($\Delta E(DEF)_B$)). The interaction energy term, $\Delta E(INT)_{A(BC)}$, is the interaction energy between A and (BC) for their respective optimized product structures.

Advanced EDA unites the natural orbitals for chemical valence (NOCV), so it is possible to separate the total orbital interactions into pairwise contributions [23]. The advanced EDA (i.e., EDA-NOCV) further divides the interaction energy ($\Delta E(INT)$) into three main components: $\Delta E(INT) = \Delta E_{elstat} + \Delta E_{Pauli} + \Delta E_{orb}$. It is used for a quantitative study of π back-bonding to fullerene ligands that uses the M06/TZ2P level of theory with the ADF 2016 program package [24]. The relativistic effect is considered by applying a scalar zero-order regular approximation (ZORA) [25]. The interaction energy and its decomposition terms are obtained from a single-point calculation using the M06/TZ2P basis set from the Gaussian 09 optimized geometry.

3. Results and discussion

3.1. Geometric changes

The structures of $\{\eta^2\text{-}(X@C_n)\}ML_2$ complexes for M = Pt, Pd; X = F^-, 0, Li^+ and n = 60, 70, 76, 84, 90 and 96 were fully optimized at the M06/LANL2DZ level of theory. The geometries that are obtained are illustrated in **Figure 1**. The key structural parameters of the stationary points are listed in **Table 1** (the structural parameters for n = 70, 76, 84, 90 and 96 are presented elsewhere). For the Pt-C_{60} complex in the absence of encapsulated ions, the respective lengths of the metal-carbon bonds are 2.12 Å and 2.12 Å (**Table 1**). When the Li^+ ion is encapsulated into the cage, the metal-carbon bonds remain unaltered and the respective distances between

http://dx.doi.org/10.5772/intechopen.70068

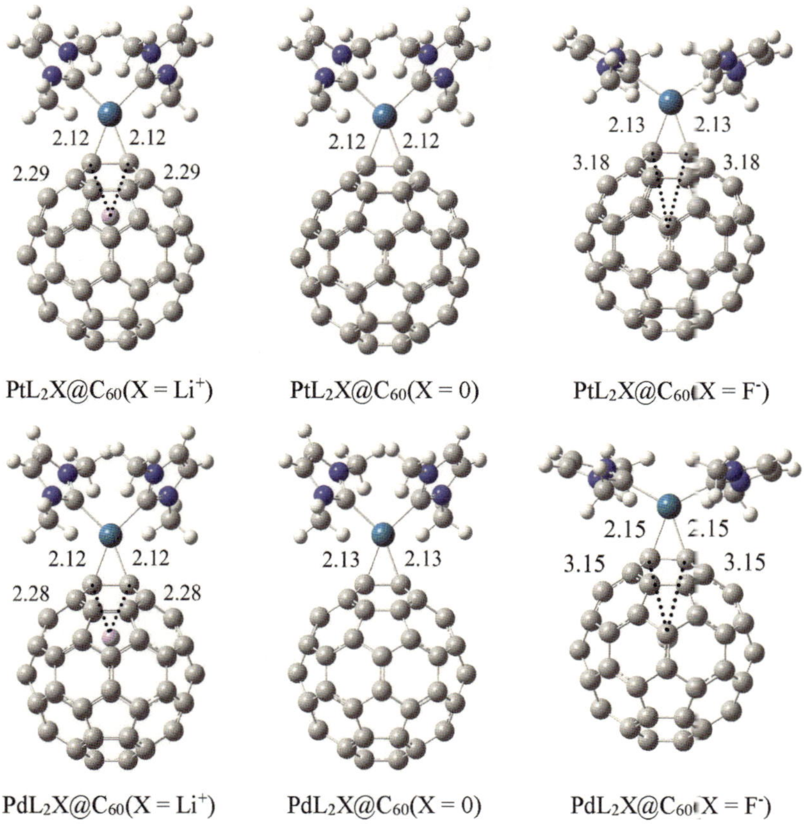

Figure 1. Optimized geometries for $\{\eta^2\text{-}(X@C_{60})\}ML_2$ (M = Pt, Pd; X = Li^+, 0, F^-).

C_1, C_2 and Li^+ are 2.29 Å and 2.29 Å. As the encapsulated ion is changed to F^-, the metal-carbon bonds remain almost unchanged (2.13 and 2.13 Å), but the distance between C_1, C_2 and encapsulated ions (F^-) increases (3.18 and 3.18 Å). The Li^+ ion is located at a site that is close to the transition metals because of electrostatic interaction. The metal-coordinated carbon atoms of C_{60} are negatively charged because there is π back-donation from the metal center. For the Pt-C_{60} complex without encapsulated ions, the natural population analysis (NPA) shows that the atomic charges on the C_1 (C_2) atoms are −0.27 (−0.27). When the cage is encapsulated by a Li^+ ion, the atomic charges on the C_1 (C_2) atoms are increased to −0.32 (−0.32) and the atomic charge on the Li atom is +0.86. Therefore, the encaged Li^+ ion is attracted toward these negatively charged C atoms. However, as the encapsulated ion is changed to F^-, NPA shows that the atomic charges on C_1 (C_2) atoms are decreased to −0.23 (−0.23) and the atomic charge on the F atom is negative (−0.93), so the encaged F^- ion is repelled by the negatively charged C atoms. In terms of Pd-C_{60} complexes, it is worthy of note that the geometrical distances are generally similar to the corresponding distances for Pt-C_{60} complexes, but the charge distributions are different. Specifically, the encaged Li atom has a charge (+0.86) but the charges on C_1 (C_2) atoms are reduced to −0.27 (−0.27). The negative charges on metal-coordinated carbon atoms are also less for X = 0 and F^-. Similar geometric changes and charge distributions are seen for n = 70, 76, 84, 90 and 96 and are presented elsewhere.

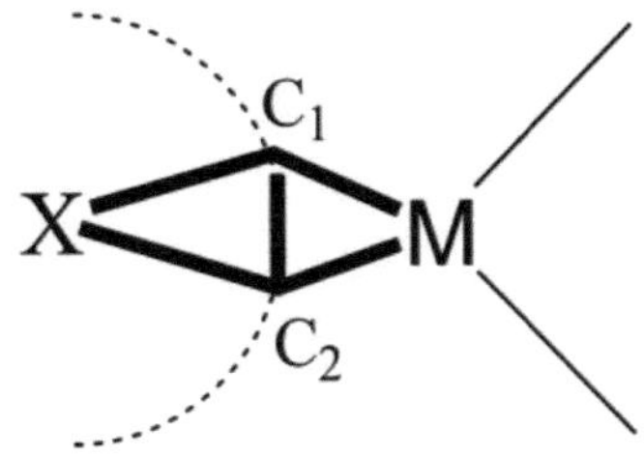

System	Geometrical parameters				NPA atomic charge			
	M-C_1	M-C_2	X-C_1	X-C_2	M	C_1	C_2	X
M = Pt; X = Li^+								
$ML_2X@C_{60}$	2.12	2.12	2.29	2.29	+0.48	−0.32	−0.32	+0.86
M = Pt; X = 0								
$ML_2X@C_{60}$	2.12	2.12	–	–	+0.46	−0.27	−0.27	–
M = Pt; X = F^-								
$ML_2X@C_{60}$	2.13	2.13	3.18	3.18	+0.43	−0.23	−0.23	−0.93
M = Pd; X = Li^+								
$ML_2X@C_{60}$	2.12	2.12	2.28	2.28	+0.44	−0.27	−0.27	+0.86
M = Pd; X = 0								
$ML_2X@C_{60}$	2.13	2.13	–	–	+0.40	−0.21	−0.21	–
M = Pd; X = F^-								
$ML_2X@C_{60}$	2.15	2.15	3.15	3.15	+0.34	−0.16	−0.16	−0.93

Table 1. Selected geometrical parameters (bond distances in Å) and the NPA atomic charge for optimized complexes ($\{\eta^2$-$(X@C_{60})\}ML_2$) at the M06/LANL2DZ level of theory.

3.2. Basic energy decomposition analysis (basic EDA)

In order to better understand the factors that govern the thermodynamic stability of $\{\eta^2$-$(X@C_n)\}ML_2$ complexes, basic EDA is performed on $\{\eta^2$-$(X@C_n)\}ML_2$ complexes (the basic EDA results for n = 70, 76, 84, 90 and 96 are presented elsewhere). The bonding energy (ΔE) is defined as $\Delta E = E(\{\eta^2\text{-}(X@C_n)\}ML_2) - E(X@C_n) - E(ML_2)$, using Eq. (1). The Pt-$C_{60}$ complex without encapsulated ions is initially considered. Basic EDA shows that both the metal fragment and the empty C_{60} fragment are distorted during the formation of the metal-carbon bond (**Table 2**). The metal fragment undergoes greater distortion ($\Delta E(DEF)_A$ = 52.1 kcal/mol) than C_{60} ($\Delta E(DEF)_B$ = 13.9 kcal/mol). The same results are obtained for X = Li^+ or F^-. It is found that the encapsulation of the Li^+ ion induces more distortion in fragments A and B of the Pt-$Li^+@C_{60}$ complex than those of the Pt-$F^-@C_{60}$ complex ($\Delta\Delta E(DEF)(X = Li^+)$ = 8.0, $\Delta\Delta E(DEF)(X = F^-)$ = −14.8 kcal/mol). However, the interaction energy increases when the Li^+ ion is encapsulated ($\Delta\Delta E(INT)_{A(BC)}(X = Li^+)$ = −35.9, $\Delta\Delta E(INT)_{A(BC)}(X = F^-)$ = +33.8 kcal/mol), which shows that the encapsulated Li^+ ion induces a stronger interaction between the metal fragment and the $X@C_{60}$ fragment, so $\{\eta^2$-$(Li^+@C_{60})\}PtL_2$

X	ΔE(DEF) (ΔE(DEF)$_A$, ΔE(DEF)$_B$)	ΔΔE(DEF)[c]	ΔE(INT)$_{A(BC)}$	ΔΔE(INT)$_{A(BC)}$[c]	ΔE[d]	ΔΔE[c,d]
{η^2-(X@C_{60})}PtL_2						
F^-	51.2 (41.0, 10.2)	−14.8	−67.0	+33.8	−14.7	+20.1
0	66.0 (52.1, 13.9)	–	−100.8	–	−34.8	–
Li^+	74.0 (55.0, 19.0)	8.0	−136.6	−35.9	−64.2	−29.4
{η^2-(X@C_{60})}PdL_2						
F^-	30.7 (25.9, 4.8)	−12.6	−45.8	29.0	−13.9	+17.6
0	43.3 (35.0, 8.3)	–	−74.8	–	−31.5	–
Li^+	51.1 (37.9, 13.2)	7.8	−109.3	−34.5	−59.2	−27.7

[a]Energies are given in kcal/mol.
[b]A and B respectively represent the metal fragment and C_{60} cage.
[c]The difference is relative to corresponding quantity at X = 0.
[d]The reaction energy without zero-point energy (ZPE) correction for the product, relative to the corresponding reactants.

Table 2. Basic EDA for {η^2-(X@C_{60})}ML_2 (M = Pt, Pd) at M06/LANL2DZ[a,b].

is more stable. The relative thermodynamic stability increases in the order: ΔE(X = F^-) < ΔE (X = 0) < ΔE(X = Li^+), as shown in **Table 2**. In addition, |ΔΔE(DEF)| is small and |ΔΔE(INT)$_{A(BC)}$| is large, so the latter must be responsible for an increase in thermodynamic stability. Similar results are obtained for Pd-C_{60} complexes, but, the distortion in fragments A or B is smaller than that for the Pt-C_{60} complex, as is the interaction energy, so the complex is less stable. For instance, ΔE(M = Pd, X = Li^+) = −34.5 > ΔE(M = Pt, X = Li^+) = −35.9 kcal/mol. When the cage size increases (n = 70, 76, 84, 90 and 96), the encapsulated Li^+ ion still induces a stronger interaction between the metal fragment and the X@C_n fragment and |ΔΔE(DEF)| is smaller than |ΔΔE(INT)$_{A(BC)}$|. Therefore, an increase in the cage size has no effect on the basic EDA results.

3.3. Advanced energy decomposition analysis (advanced EDA)

In an earlier study by the authors, structural parameters and spectral characteristics were used to estimate the strength of π back-bonding for {η^2-(X@C_{60})}ML_2 (M = Pt, Pd; X = 0, Li^+, L = PPh_3) complexes [10]. The changes in bond length ($\Delta r/r_0$), bond angle ($\Delta \vartheta_{av}$), vibrational frequency (Δv) and the chemical shift (Δδ) were used to describe the character of the π-complex s. In this study, the strength of the π back-bonding strength is estimated from an energetic viewpoint using an advanced EDA method. This analysis shows the effect of encapsulated ions, metal fragments and cage sizes on π back-bonding.

3.3.1. *The effect of encapsulated ions on π back-bonding*

In an earlier discussion (Section 3.2), it was proven that thermodynamic stabilities increase in the order: ΔE(X = F^-) < ΔE(X = 0) < ΔE(X = Li^+), because the interaction energy (ΔE(INT)) is increased. The interaction between the metal fragment and $X@C_n$ is now studied using advanced EDA, which further decomposes the interaction energy into electrostatic interaction (ΔE_{elstat}), repulsive Pauli interaction (ΔE_{Pauli}) and orbital interaction (ΔE_{orb}) terms. The orbital interactions are the most important of these three and only the most important pairwise contributions to ΔE_{orb} are considered. The advanced EDA method is used for $\{\eta^2\text{-}(X@C_n)\}ML_2$ complexes, as shown in **Tables 3** and **4** (the results for n = 70, 76, 84, 90 and 96 are presented elsewhere). A plot of the deformation density and a qualitative drawing of the orbital interactions between the metal fragment and $X@C_{60}$ are shown in **Figure 2**. In terms of the Pt-C_{60} complexes, **Table 3** shows that both the electrostatic interaction (ΔE_{elstat}) and the orbital interaction (ΔE_{orb}) stabilize the complexes because they are negative terms, but the percentage of ΔE_{orb} increases in the order: $\Delta E_{orb}(X = F^-) < \Delta E_{orb}(X = 0) < \Delta E_{orb}(X = Li^+)$. Therefore, the enhanced orbital interaction must be responsible for the increase in the thermodynamic stability. **Table 3** also shows that ΔE_1 contributes significantly to ΔE_{orb}: 69.5% for $X = F^-$, 75.2% for X = 0 and 76.4% for $X = Li^+$. The deformation densities show that these come from π back-donation from a filled *d* orbital of the metal to the π^* orbitals of C_{60} (charge flow is yellow to green at the top of **Figure 2c**). The large contributions of ΔE_1 to ΔE_{orb} are in agreement with the results of previous studies. The metal-carbon bonds are principally formed by π back-donation [8]. It is also seen that the order of ΔE_1 is $|\Delta E_1(X = F^-)| = 94.4 < |\Delta E_1(X = 0)| = 118.6 < |\Delta E_1(X = Li^+)| = 142.8$ kcal/mol. Therefore, ΔE_1 is increased when there is the encapsulated Li^+ ion but decreased when there is a F^- ion. The second contribution of ΔE_2 to ΔE_{orb} is comparatively small: 18.0% for $X = F^-$, 13.1% for X = 0 and 10.0% for $X = Li^+$. This results from

Fragments	**L_2Pt and $F^-@C_{60}$**	**L_2Pt and C_{60}**	**L_2Pt and $Li^+@C_{60}$**
ΔE_{int}	−67.7	−100.8	−133.0
ΔE_{Pauli}	257.2	235.8	227.6
ΔE_{elstat} [b]	−189.0 (58.2%)	−178.8 (53.1%)	−173.8 (48.2%)
ΔE_{orb} [b]	−135.8 (41.8%)	−157.7 (46.9%)	−186.9 (51.8%)
ΔE_1 [c]	−94.4 (69.5%)	−118.6 (75.2%)	−142.8 (76.4%)
ΔE_2 [c]	−24.4 (18.0%)	−20.6 (13.1%)	−18.6 (10.0%)
ΔE_3 [c]	−7.1 (5.2%)	−5.7 (3.6%)	−6.5 (3.5%)
ΔE_4 [c]	−3.0 (2.2%)	−3.8 (2.4%)	−5.5 (3.0%)
ΔE_5 [c]	−3.6 (2.7%)	−4.2 (2.7%)	−5.0 (2.7%)
ΔE_6 [c]	–	–	−2.1 (1.1%)
ΔE_{rest} [c]	−5.8 (4.3%)	−6.3 (4.0%)	−6.8 (3.6%)

[a]Optimized structures at the M06/LANL2DZ level of theory.
[b]The values in parentheses give the percentage contribution to the total attractive interactions, $\Delta E_{elstat} + \Delta E_{orb}$.
[c]The values in parentheses give the percentage contribution to the total orbital interactions, ΔE_{orb}.

Table 3. The advanced EDA results for $\{\eta^2\text{-}(X@C_{60})\}PtL_2$[a] ($X = F^-$, 0, Li^+) at the M06/TZ2P level of theory. The fragments are PtL_2 and $X@C_{60}$ in a singlet (S) electronic state. All energy values are in kcal/mol.

http://dx.doi.org/10.5772/intechopen.70068

Fragments	L_2Pd and $F^-@C_{60}$	L_2Pd and C_{60}	L_2Pd and $Li^+@C_{60}$
ΔE_{int}	−45.1	−73.5	−103.9
ΔE_{Pauli}	188.9	176.5	176.4
ΔE_{elstat}[b]	−142.8 (61.0%)	−137.0 (54.8%)	−138.9 (49.6%)
ΔE_{orb}[b]	−91.2 (39.0%)	−113.0 (45.2%)	−141.3 (50.4%)
ΔE_1[c]	−71.9 (78.8%)	−96.9 (85.8%)	−121.0 (85.6%)
ΔE_2[c]	−15.5 (17.0%)	−10.8 (9.6%)	−9.4 (6.7%)
ΔE_3[c]	−4.9 (5.4%)	−4.2 (3.7%)	−4.6 (3.3%)
ΔE_4[c]	−2.1 (2.3%)	−2.3 (2.0%)	−3.6 (2.5%)
ΔE_5[c]	−2.0 (2.2%)	−2.9 (2.6%)	−3.6 (2.5%)
ΔE_{rest}[c]	−2.7 (3.0%)	−3.6 (3.2%)	−6.5 (4.6%)

[a]Optimized structures at the M06/LANL2DZ level of theory.
[b]The values in parentheses give the percentage contribution to the total attractive interactions $\Delta E_{elstat} + \Delta E_{orb}$.
[c]The values in parentheses give the percentage contribution to the total orbital interactions, ΔE_{orb}.

Table 4. The advanced EDA results for $\{\eta^2\text{-}(X@C_{60})\}PdL_2$[a] ($X = F^-$, 0, Li^+) at the M06/TZ2P level of theory. The fragments are PdL_2 and $X@C_{60}$ in the singlet (S) electronic state. All energy values are in kcal/mol.

σ-donation from a filled π orbital of C_{60} to the π* orbital of the metal (middle of **Figure 2c**). The computational results show that π back-bonding is crucial to the thermodynamic stability of Pt-C_{60} complexes and that an encapsulated Li^+ ion increases π back-bonding but an encapsulated F^- has the opposite effect.

3.3.2. *The effect of metal fragments on π back-bonding*

Pd-C_{60} complexes appear to be similar to Pt-C_{60} complexes, but a comparison of the results in **Tables 3** and **4** shows that the value ΔE_1 for a Pd-C_{60} complex is smaller than the corresponding value for a Pt-C_{60} complex, which demonstrates that the π back-bonding for a palladium center is weaker than that for a platinum center. For example, $|\Delta E_1(M = Pd, X = Li^+)| = 121.0 < |\Delta E_1(M = Pt, X = Li^+)| = 142.8$ kcal/mol. This is consistent with the earlier results that were obtained using structural parameters and spectral characteristics [10].

3.3.3. *The effect of cage sizes on π back-bonding*

Figure 3 shows a plot of the ΔE_1 values versus cage sizes that are calculated for $\{\eta^2\text{-}(X@C_n)\}PtL_2$ complexes (n = 60, 70, 76, 84, 90 and 96). It is seen that there is no linear relationship and there is one obvious peak for each X at n = 84 [26]. This demonstrates the effect of a difference in size of the carbon clusters on π back-bonding for a metal center, but the correlation is not simply monotonic. Therefore, a larger (smaller) cage size does not necessarily imply that there is stronger (weaker) π back-bonding, which results in greater (lower) thermodynamic stability.

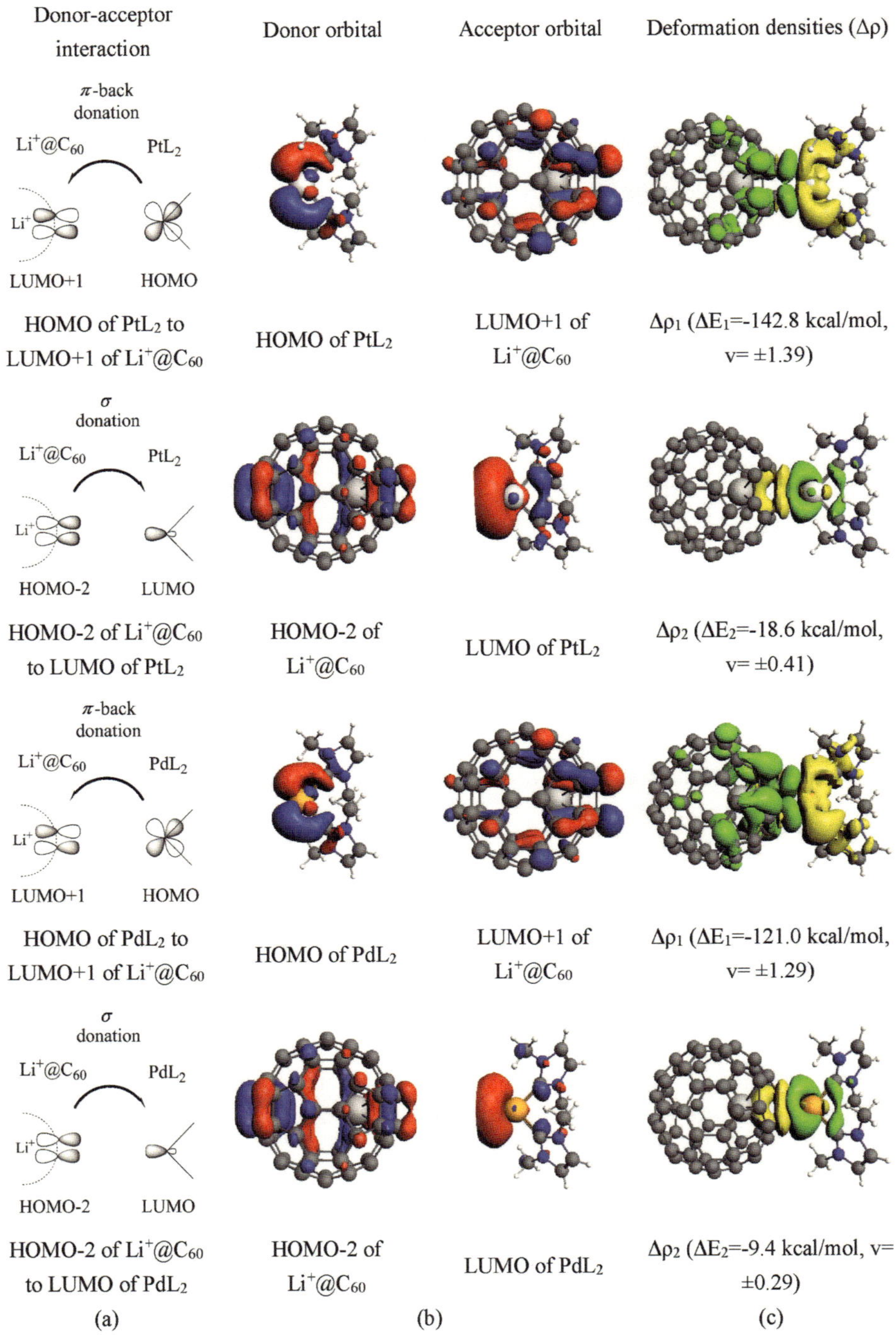

Figure 2. (a) A qualitative drawing of the orbital interactions between the metal fragment and $Li^{+}@C_{60}$; (b) the shape of the most important interacting occupied and vacant orbitals of the metal fragments and $Li^{+}@C_{60}$; (c) a plot of the deformation densities, Δϱ, for the pairwise orbital interactions between the two fragment in their closed-shell state, the associated interaction energies, ΔE_{orb} (in kcal/mol), and the eigenvalues ν. The eigenvalues, ν, indicate the size of the charge flow. The direction of the charge flow is from yellow to the green.

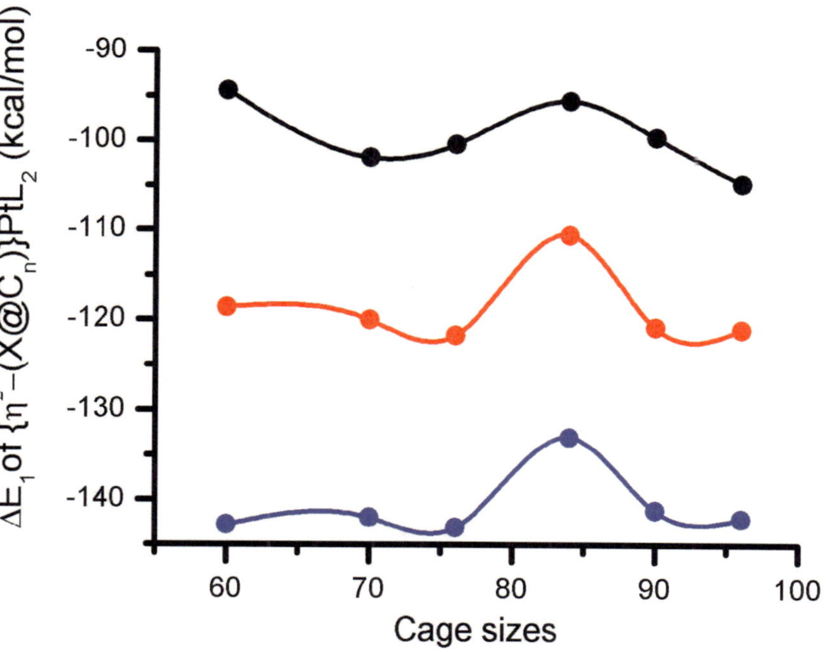

Figure 3. The correlation between ΔE_1 and cage sizes for $\{\eta^2\text{-}(X@C_n)\}PtL_2$ (n = 60, 70, 76, 84, 90 and 96) complexes. The blue, red and black lines indicate the ΔE_1 for X = Li^+, 0 and F^-, respectively.

4. Conclusion

This computational study uses density functional theory to determine the thermodynamic stability of $\{\eta^2\text{-}(X@C_n)\}ML_2$ complexes (M = Pt, Pd; X = F^-, 0, Li^+ and n = 60, 70, 76, 84, 90 and 96). The calculations show the reaction is more stable when the Li^+ ion is encapsulated within C_n but the complex becomes unstable if there is a F^- ion. Basic EDA shows that there is an increase in the interaction between the metal fragment and C_n if there is an encapsulated Li^+ ion but F^- ion has the opposite effect.

The advanced EDA results show that π back-bonding is crucial to thermodynamic stability and that thermodynamic stability is increased by the presence of a Li^+ ion but the presence of a F^- ion has the opposite effect. These computations also show that a platinum center results in stronger π back-bonding than a palladium center and that there is no linear relationship between cage size and π back-bonding.

Acknowledgements

The authors would like to thank the National Center for High-Performance Computing in Taiwan for the donation of generous amounts of computing time. The authors are also grateful for financial support from the Ministry of Science and Technology of Taiwan.

Author details

Ming-Chung Yang[1] and Ming-Der Su[1,2*]

*Address all correspondence to: midesu@mail.ncyu.edu.tw

1 Department of Applied Chemistry, National Chiayi University, Chiayi, Taiwan

2 Department of Medicinal and Applied Chemistry, Kaohsiung Medical University, Kaohsiung, Taiwan

References

[1] Fagan PJ, Calabrese JC, Malone B. The chemical nature of buckminsterfullerene (C_{60}) and the characterization of a platinum derivative. Science. 1991;**252**:1160-1161. DOI: 10.1126/science.252.5009.1160

[2] Lebedeva MA, Chamberlain TW, Khlobystov AN. Harnessing the synergistic and complementary properties of fullerene and transition-metal compounds for nanomaterial applications. Chemical Reviews. 2015;**115**:11301-11351. DOI: 10.1021/acs.chemrev.5b00005

[3] Balch AL, Catalano VJ, Lee JW. Accumulating evidence for the selective reactivity of the 6-6 ring fusion of C_{60}. Preparation and structure of $(\eta^2\text{-}C_{60})Ir(CO)Cl(PPh_3)_2 \cdot 5C_6H_6$. Inorganic Chemistry. 1991;**30**:3980-3981. DOI: 10.1021/ic00021a003

[4] Balch AL, Costa DA, Noll BC, Olmstead MM. Oxidation of buckminsterfullerene with m-chloroperoxybenzoic acid. Characterization of a Cs isomer of the diepoxide $C_{60}O_2$. Journal of the American Chemical Society. 1995;**117**:8926-8932. DOI: 10.1021/ja00140a005

[5] Balch AL, Winkler K. Two-component polymeric materials of fullerenes and the transition metal complexes: A bridge between metal–organic frameworks and conducting polymers. Chemical Reviews. 2016;**116**:3812-3882. DOI: 10.1021/acs.chemrev.5b00553

[6] Tsikalova MV, Zheludkov SV, Vorontsov EI, Bashilov VV, Babievskii KK, Sokolov VI, Novikov YN. Optically active rhodium and iridium C_{60} complexes containing the enantiomeric ligand (+)DIOP: $(\eta^2\text{-}C_{60})MH(CO)[(+)DIOP]$ (M = Rh, Ir). Mendeleev Communications. 2011;**21**:256. DOI: 10.1016/j.mencom.2011.09.008

[7] Proft FD, Alsenoy CV, Geerlings P. Ab initio study of the endohedral complexes of C_{60}, Si_{60}, and Ge_{60} with monoatomic ions: Influence of electrostatic effects and hardness. The Journal of Physical Chemistry. 1996;**100**:7440-7448. DOI: 10.1021/jp960174w

[8] Watanabe T, Itoh MF, Komuro T, Okada H, Sakai T, Ono Y, Kawachi K, Kasama Y, Tobita H. Iridium and platinum complexes of $Li^+@C_{60}$. Organometallics. 2014;**33**:608-611. DOI: 10.1021/om4008899

[9] Simoes JAM, Beauchamp JL. Transition metal-hydrogen and metal-carbon bond strengths: The keys to catalysis. Chemical Reviews. 1990;**90**:629-688. DOI: 10.1021/cr00102a004

[10] Yang MC, Sharma AK, Sameera WMC, Morokuma K, Su MD. A theoretical study of addition reactions of $L_4M(M = Rh, Ir)$ and $L_2M(M = Pd, Pt)$ to $Li^+@C_{60}$. The Journal of Physical Chemistry A. 2017;**121**:2665-2673. DOI: 10.1021/acs.jpca.7b01086

[11] Hopkinson MN, Richter C, Schedler M, Glorius F. An overview of N-heterocyclic carbenes. Nature. 2014;**510**:485-496. DOI: 10.1038/nature13384

[12] Balch AL, Olmstead MM. Reactions of transition metal complexes with fullerenes (C_{60}, C_{70}, etc.) and related materials. Chemical Reviews. 1998;**98**:2123-2165. DOI: 10.1021/cr960040e

[13] Yang H, Beavers CM, Wang Z, Jiang A, Liu Z, Jin H, Mercado BQ, Olmstead MM, Balch AL. Isolation of a small carbon nanotube: The surprising appearance of $D_{5h}(1)$-C_{90}. Angewandte Chemie International Edition. 2009;**48**:1-6. DOI: 10.1002/anie.200906023

[14] Yang H, Jin H, Che Y, Hong B, Liu Z, Gharamaleki JA, Olmstead MM, Balch AL. Isolation of four isomers of C_{96} and crystallographic characterization of nanotubular $D_{3d}(3)$-C_{96} and the somewhat flat-sided sphere $C_2(181)$-C_{96}. Chemistry—A European Journal. 2012;**18**:2792-2796. DOI: 10.1002/chem.201103852

[15] Fowler PW, Manolopoulos DE. An Atlas of Fullerenes. Oxford: Clarendon; 1995. DOI: http://dx.doi.org/10.1080/10641229608001575

[16] Hagebaum-Reignier D, Girardi R, Carissan Y, Humbel S. Hückel theory for Lewis structures: Hückel–Lewis Configuration Interaction (HL-CI). Journal of Molecular Structure: THEOCHEM. 2007;**817**:99-109. DOI: 10.1016/j.theochem.2007.04.026

[17] Zhao Y, Truhlar DG. The M06 suite of density functionals for main group thermochemistry, thermochemical kinetics, noncovalent interactions, excited states, and transition elements: Two new functionals and systematic testing of four M06-class functionals and 12 other functionals. Theoretical Chemistry Accounts. 2008;**120**:215-241. DOI: 10.1007/s00214-007-0310-x

[18] Dunning TH, Hay PJ. In: Schaefer HF, editor. Modern Theoretical Chemistry. Vol. 3. New York: Plenum; 1977. pp. 1-28

[19] Hay PJ, Wadt WR. Ab initio effective core potentials for molecular calculations—Potentials for the transition-metal atoms Sc to Hg. The Journal of Chemical Physics. 1985;**82**:270-283. DOI: http://dx.doi.org/10.1063/1.448799

[20] Frisch MJ, Trucks GW, Schlegel HB, Scuseria GE, Robb MA, Cheeseman JR, Scalmani G, Barone V, Mennucci B, Petersson GA, et al. Wallingford, CT: Gaussian, Inc.; 2013

[21] Sameera WMC, Hatanaka M, Kitanosono T, Kobayashi S, Morokuma K. The mechanism of iron(II)-catalyzed asymmetric Mukaiyama aldol reaction in aqueous media: Density functional theory and artificial force-induced reaction study. Journal of the American Chemical Society. 2015;**137**:11085-11094. DOI: 10.1021/jacs.5b05835

[22] Ziegler T, Rauk A. Carbon monoxide, carbon monosulfide, molecular nitrogen, phosphorus trifluoride, and methyl isocyanide as σ donors and π acceptors. A theoretical study by the Hartree-Fock-Slater transition-state method. Inorganic Chemistry. 1979;**18**:1755-1759. DOI: 10.1021/ic50197a006

[23] Zhao L, Jones C, Frenking G. Reaction mechanism of the symmetry-forbidden [2+2] addition of ethylene and acetylene to amido-substituted digermynes and distannyes Ph_2N-EE-NPh_2, (E = Ge, Sn). A theoretical study. Chemistry–A European Journal. 2015;**21**:12405-12413. DOI: 10.1002/chem.201501457

[24] Bickelhaupt FM, Baerends EJ. Kohn-Sham density functional theory: Predicting and understanding chemistry. Reviews in Computational Chemistry. Wiley-VCH, New York. 2000;**15**:1-86

[25] Lenthe EV, Baerends EJ, Snijders JG. Relativistic regular two-component Hamiltonians. The Journal of Chemical Physics. 1993;**99**:4597-4610. DOI: http://dx.doi.org/10.1063/1.466059

[26] The reason for such a phenomenon is still unclear. Presumably, it would be because of the smallest Hückel π bond order (B = 0.521) for n = 84. The small π bond order is expected to be inactive, thus, leading to weak π back-bonding interaction. Further calculations are working on and will be present elsewhere

Section 3

On Application Aspects

Chapter 8

Fullerene as Spin Converter

Elif Okutan

Additional information is available at the end of the chapter

http://dx.doi.org/10.5772/intechopen.74541

Abstract

It has now been more than 30 years since buckminsterfullerene became a real thing. An exclusive field of study called fullerene chemistry arises and is discovered to be unique in many respects. In a very short time, a great deal of effort has been devoted on the fullerene chemistry and properties of fullerene derivatives. The fields of fullerene and light-induced processes considerably overlapped, and now numerous wonderful examples are on exhibition. The large number of these systems has been designed to take advantage of the electron-accepting property of fullerene and broadcasts the fullerenes as universal spin converter advertising its perfect intersystem crossing (ISC) quantum yield where a spin converter can be identified as a chromophore that undergoes efficient ISC with a low first excited state (S_1), but does not necessarily strongly harvest visible light. Thus, donor-acceptor systems in the field of light-induced processes within multicomponent fullerene arrays have been proposed as models for optical limiters, owing to the singlet oxygen production efficiency of C_{60} and the C_{60} derivatives in the field of medicinal chemistry.

Keywords: fullerene, spin converter, light harvesting, singlet oxygen, triplet-triplet annihilation

1. Introduction

Transformation of known and the creation of new are always intrigued to synthetic chemists. Not long ago, elemental carbon was not even a figurant as starting material. This position promoted fiercely after the family of carbon allotropes enlarged by the welcoming new members to the core family "graphite and diamond." Unlike to graphite and diamond, fullerenes are spherical molecules with solid-state structure and are soluble in various solvents that opened a new era for chemical manipulation of carbon-based materials [1].

Few discoveries like this are captured the attention of scientist and the general public alike as much as the discovery of these architecturally esthetic molecules. But the popularity of fullerenes in science is not merely due to esthetics. Years of intense research activity showed that C_{60} is a powerful building block to be used in materials science and medicinal chemistry [2, 3]. What put fullerenes in the heart of nanotechnology today is the association of several extreme properties, such as outstanding mechanical, thermal, electronic, and electrical properties, coupled with chemical robustness, which have spurred a broad range of applications that provides new research possibilities for scientists, particularly in terms of electron-acceptor proficiency, both in the solid state and in solution [4]. An entirely new discipline called as "fullerene chemistry" emerged [5–7].

2. Optical studies of fullerene

Fullerene structure facilitates ingrained synthetic methodologies that catalyzing the production of a wide variety of novel derivatives often encompasses many fields outside the traditional scope [8, 9]. One of the most exciting properties of fullerene chemistry is related to their excited-state properties [10, 11]. The most abundant representative of the family C_{60} can be identified via its strong absorptions between 190 and 410 nm (allowed $^1T_{1u}$–^1A_g) as well as by some pale but significant transitions in the 410 and 620 nm (orbital forbidden singlet-singlet) region which is responsible for the purple color of C_{60} and the red color of C_{70} [12, 13].

The fullerenes, in particular C_{60}, exhibit a variety of remarkable photophysical properties, making them very attractive building blocks for the construction of photosynthetic antenna and reaction center models that result from their large pi electron system that cater dense manifolds of low-lying electronically excited states [14–17]. As it is given below, most photochemical and photophysical applications of fullerenes are likely be mediated by lowest of these energetic states which present triplet spin multiplicity (**Figure 1**).

The fullerenes can also be identified as useful optical limiters since the triplet-triplet absorptivities are higher than the ground-state absorptions [20, 21]. The singlet excited state (1.99 eV) of C_{60} efficiently decays to the lower lying triplet excited state (1.57 eV) via intersystem crossing [17]. The triplet quantum yields are very high. The triplet excited states are responsive to diverse processes for deactivation, such as ground-state quenching, triplet-triplet annihilation via molecular oxygen leading to $^1O_2$, and electron transfer to donor molecules [17]. The long-lived triplet states due to the ISC process of excited singlet states gave rise to a substantial interest for their prosperous applications, such as singlet oxygen generation for [22–26], enhancing photoinduced electric conductivity via formation of charge-separated state caused by the excited triplet state of fullerene [27–29] and in the applications of photovoltaics, photocatalysis, photoinduced charge separation, and molecular probes (**Figure 2**) [30–36].

In this account, we discuss some of the main achievements in this rapidly developing field and, in particular, the triplet photosensitizer (PS) phenomena associated with the excited triplet state of fullerenes that later used as catalysts in photochemical reactions. Triplet PSs as the name derived from the compounds, used to transfer energy to other, are used for not only

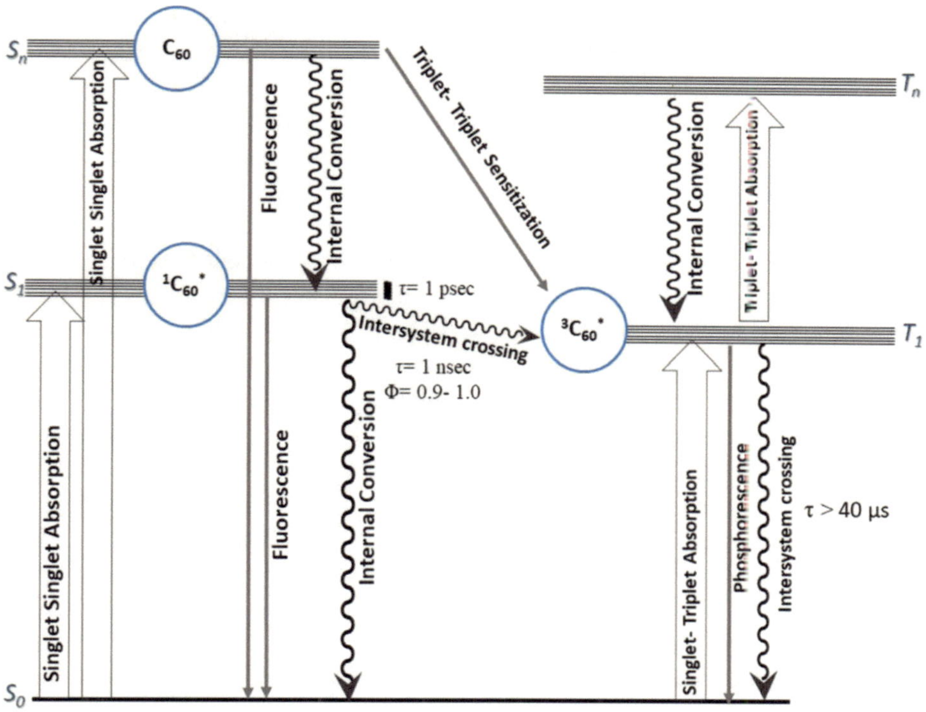

Figure 1. The Jablonski diagram: schematic depiction of the energy levels of typical compound (adapted from [18, 19]).

$$C_{60}\,(T_1) \xrightarrow{k_1^{60}} C_{60}\,(S_0) \quad \text{intersystem crossing} \quad \mathbf{(1)}$$

$$C_{60}\,(T_1) + O_2 \xrightarrow{k_{O_2}^{60}} C_{60}\,(S_0) + O_2 \quad \text{oxygen quenching} \quad \mathbf{(2)}$$

$$C_{60}\,(T_1) + C_{60}\,(S_0) \xrightarrow{k_{aq}^{60}} 2\,C_{60}\,(S_0) \quad \text{self- quenching} \quad \mathbf{(3)}$$

$$2\,C_{60}\,(T_1) \xrightarrow{k_{T-T}^{60}} C_{60}\,(S_0) + C_{60}\,(T_1) \quad \text{triplet- triplet annihilation} \quad \mathbf{(4)}$$

Figure 2. Main decay channels that control the lifetime of triplet C_{60} in solution [37]. (reprinted by permission of ACS Publications).

energy transfer but also for photovoltaic reactions such as photodynamic therapy (PDT), photo-induced hydrogen production from water, and triplet-triplet annihilation (TTA) upconversion systems. The photosensitizing properties of some relevant classes of functionalized fullerene-based materials are surveyed.

Even in the early reports on excited states of fullerene prepared via light excitation, researchers were concerned with the persistence of long-lived excited and low triplet states. A number of researches addressed the decay kinetics of fullerenes (C_{60} and C_{70}), and the unimolecular triplet state lifetimes are found to be significantly longer than recognized [22, 38–45]. In the bimolecular triplet state processes, there are various deactivation mechanisms including oxygen quenching, triplet-triplet annihilation, triplet energy exchange, and self-quenching. The deactivation of organic triplet states by dissolved oxygen is reported by numerous researchers, and oxygen quenching constants for C_{60} and C_{70} are recorded in room temperature [22, 23] and were persistent with previous results [37]. Triplet-triplet annihilation occurs between two clashing triplet states considering the deactivation of one molecule, while exposure of the other in a highly excited state that will quickly convert to first singlet excited state (S_1) (**Figure 2**). The reported kTT values are 5.4 x 10^9 M^{-1} s^{-1} and 5.5 x 10^9 M^{-1} s^{-1} in room temperature which should considered to be high and kept in mind to eliminate contributions from this channel while measuring excited state kinetics such as studying with low concentration [37, 46]. In another scenario triplet energy may flow between two species in solution [46]. If one species T_1 state is significantly lower than the other, the second specimen will collect the energy and behave as a triplet quencher. On another scenario, the two *T1* states might be separated by several kTT. In this case, reverse energy transfer may also be occurred, and results showed that reversible energy transfer between triplet states C_{60} and C_{70} is a fast and efficient process [47]. Also, even though no mechanism was proposed for self-quenching mechanisms, the ground-state concentration dependence refers to the encounter of a molecule on T_1 state with ground-state molecule. Studies showed that fullerene displays strong self-quenching over organic triplet states [48, 49].

Fullerene exhibits unique C–C single and double bonds. The deficiency of high-energy C–H and O–H vibrations makes these materials very interesting in photonics. The usual materials such as polymers have absorption in the near infrared because of the overtones of the abovementioned vibrations. No such absorption is observed in fullerenes while exhibiting narrow electronic bands and resonances [50, 51]. Fullerene's large number of conjugated double bonds lead to large nonlinear polarizabilities. The third-order optical polarizability, γ, is always symmetry allowed, while the second-order polarizability, β, is reported to be zero for C_{60} and C_{70} since they have centrosymmetric structures [51]. By preparing charge-transfer complexes (fullerene as electron acceptor), the center of symmetry is interrupted, and second-order optical nonlinearity is induced [52]. Specifically, the optical limiting in fullerene based on the reverse saturable absorption which takes place when the excited state absorption cross section is bigger than that of the ground state. This effect was reported for C_{60} and for C_{70} under 532 nm and 1064 nm excitations where optical limiting performance of C_{60} is bigger than C_{70} since the latter exhibits a higher linear state absorption coefficient [53].

The design of new structures with high first hyperpolarizability (β) can be made via two state models in which β is expressed with the dipole moment difference, the transition dipole moment, and the energy difference between the key charge-transfer excited state and the ground electronic state [54]. The charge-transfer complexes can be characterized by the absorption cross section for an excited state of the organic moiety, which is significantly greater than that for the ground state [53]. Thus, these nanomaterials offer the optical limiting phenomena, particularly in the IR range. Fullerene-induced sensitization also favors bathochromic shift in the absorption spectra of related structures and activate transition in the near- and middle-IR range [55]. During the charge transfer, an additional electric field gradient is reported to be formed as substitute from the intramolecular donor to fullerene rather than to the intramolecular acceptor. Consequently, the nanomaterial offers high-frequency Kerr effect and exhibits high value of nonlinear refraction and nonlinear third-order permeability [56, 57]. In general, photoinduced electron transfer in donor-acceptor dyads in solution is reported to be related to free energy change for charge separation ΔG_{CS}, which depends on the energies for oxidation, reduction, and excitation and also to the Coulomb interaction and solvation of the radical ions [58]. The rate for electron transfer is obtained from the charge transfer and the electronic coupling of donor and acceptor in the excited state barrier. Also, energy transfer processes dipole-dipole (Förster) and exchange (Dexter) mechanisms are generally used to explain the deactivation of the initial photoexcited state.

Afterward, fullerene chemistry allowed researchers to open a new door to link fullerenes to photoactive species, and the work to date suggests that the first excited state can be populated by singlet-singlet energy transfer from attached dye and in the appropriate conditions can be quenched by triplet energy acceptors. Thus, fullerenes hold a significant promise, and the study on these materials will be a scientific endeavor [37, 43, 59–67].

On the account of simple organic molecules, the triplet manifold is rarely reached by direct stimulation of the organic molecule, unless advertised spin-orbit coupling effect is populated by a heavy atom [68, 69]. The triplet manifold may be reached via indirect secondary routes such as intersystem crossing from the first excited state (S_1) and charge separation and recombination between radical ions [70–73]. This kind of recombination may require orbital contact via flexible link or by a rigid spacer [74–79]. Ziessel et al. described results of photophysical properties of a closely spaced molecular dyad comprising terminal BODIPY dye and a fullerene. They investigated the driving force for light-induced electron abstraction by the S_1 state of C_{60} from BODIPY unit and displayed the dependence to solvent polarity. In nonpolar solvents, fast excitation energy transfer was declared, while electron transfer became laborious. Polar solvents play a critical role switching on light-induced electron transfer [58].

Artificial photosynthetic systems to mimic natural systems for global energy demand are important not only to understand nature but also for environmental issues. Various synthetic models were designed and constructed based on tetrapyrroles as energy harvesting antennae due to their structural resemblance to natural chlorophylls [80–83]. Owing to fullerene's facile reduction and low reorganization energy, fullerene lessened the use of 2D electron donor-acceptor systems such as quinone and methyl viologen and was successfully demonstrated in studies as "antenna-reaction center." Maligaspe et al. developed supramolecular triads to mimic these issued antenna-reaction center systems designing boron dipyrrin (BODIPY) entities as antenna that linked to zinc porphyrin (P) as electron donor and then coordinate to fullerene as electron acceptor (**Figure 3**) [84]. Similarly, BODIPY-ZnPc-fullerene system, where BODIPY unit located on peripheral position on Pc was designed, demonstrates a sequence of energy and electron transfer reactions upon photoexcitation [85]. An interesting example of distyryl-BODIPY-fullerene donor-acceptor system was also reported by Liu et al [86].

Upconversion systems are used in photovoltaics, photocatalysis, nonlinear optics, and luminescent molecular probes [87]. To facilitate upconversion one or two methods are deployed for

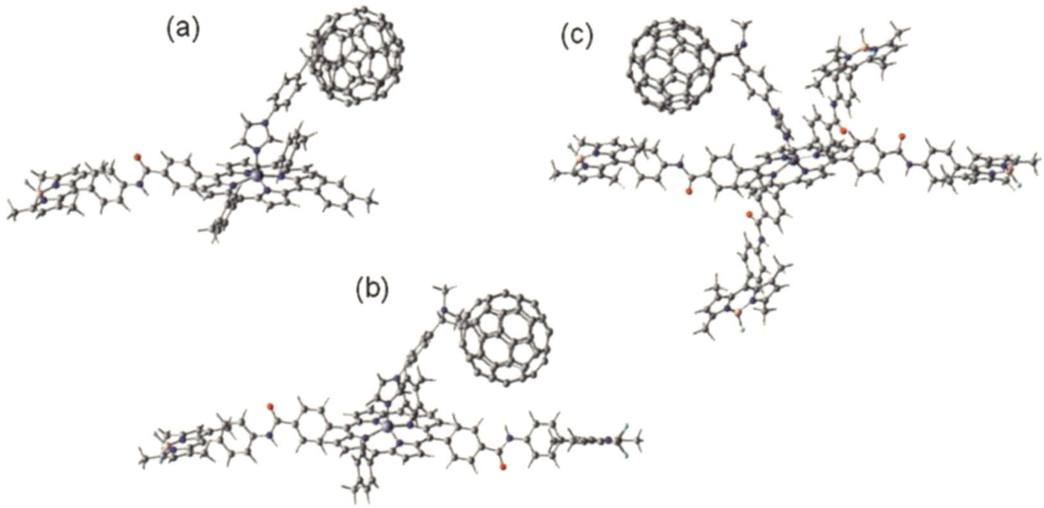

Figure 3. Optimized structures of BODIPY-ZnPc-C_{60} triads [85]. (Reprinted by permission of RSC).

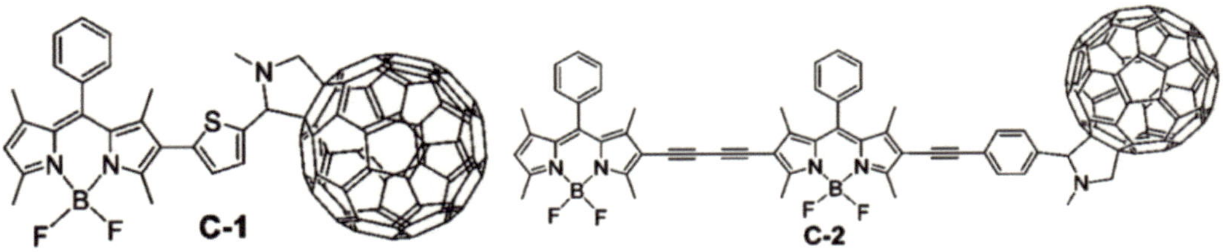

Figure 4. Structure of the C_{60}-based dyads as organic triplet sensitizers for triplet-triplet annihilation upconversions [36] (Reprinted by permission of ACS).

such as rare-earth materials or two-photon absorption fluorescent dyes [88, 89]. These conventional upconversion methods come with the disadvantages of weak absorption, low upconversion quantum yield, or requirement of consistent high-power density excitation source. New upconversion method based on triplet photosensitizer and triplet accelerator become popular where triplet photosensitizers are responsible for light harvesting and enhancement of triplet excited state by intersystem crossing (ISC) [90]. The previously reported triplet photosensitizers are usually transition-metal complexes, such as PtII/PdII porphyrin complexes, IrIII/RuII complexes, or heavy-atom derivatives of organic fluorophores since ISC is facilitated by the heavy atom spin-orbital coupling effect [91–93]. Designing these systems via chemical derivatization of a known heavy atom-free organic triplet photosensitizer is not a decisive way to prepare new organic triplet photosensitizers because even simple derivatization of the chromophore may diminish ISC [25, 26]. Heavy atom-free organic triplet photosensitizers with absorption in visible range are highly desired and remained rare for TTA upconversion system (**Figure 4**) [36, 87, 94]. An intramolecular spin converter is used to overcome the aforementioned challenges. Both C_{60} and C_{70} were also used to construct heavy atom-free triplet photosensitizer, both red-to-green and green-to-blue upconversions [36, 95, 96].

Lim et al. also reported a supramolecular tetrad bearing covalently linked ferrocene-zinc porphyrin-BODIPY system coordinated to fullerene that proposed as photosynthetic antenna reaction center mimicked by performing systematic spectral, computational, and electrochemical studies to evaluate the role of each entity in the photochemical reactions [97].

Even the early studies on evaluation of using fullerene derivatives to generate singlet oxygen. Even if there was little quantitative data at that time displayed the singlet oxygen generation by dissolving C_{60} in polyvinylpyrrolidone (mutagenic for *Salmonella* strains TA102, TA104, and YG3003) in the presence of rat liver microsomes followed by the irradiation with visible light [98]. Singlet oxygen efficiency dependence on the kind and number of addends was also studied during the early period of fullerene chemistry. The results suggested that efficiency easy independent from the kind of addends but decreased with an increasing number of the substituent [99]. A strong correlation was also reported by Prat et al. between triplet properties and the topology of the fullerene core [100]. Since fullerene chemistry evolved during that time, several fullerene derivatives were prepared to effectively generate singlet oxygen for numerous applications [101–104]. For their possible application in photodynamic therapy, a prototype macromolecule bearing a distyrylbenzene dimer as TPA unit and a [60] fullerene moiety for singlet oxygen generation endowed with a high two-photon absorption (TPA) cross section and a high singlet oxygen quantum yield were reported by Collini et al. (**Figure 5**) [105]. The singlet oxygen generation and photoinduced charge separation of zinc phthalocyanine-fullerene dyad bearing tetra polyethylene glycol moieties were also reported for PDT application [106].

In organic synthesis, oxidation is one of the primary reactions; thus, there has been an extensive research interest devoted on the use of singlet oxygen as photocatalysis in photooxidation reactions [25, 26, 107, 108]. Huang et al. used energy funneling for the first time with C_{60}-BODIPY triads and tetrads as dual functional photocatalysis for two different photocatalytic reactions. They produce juglone via photooxidation of naphthol and superoxide radical ion by photocatalytic aerobic oxidation of aromatic boronic acids to produce phenol. Reaction time was also reduced reasonably with this strategy [109]. Novel heavy atom-free triplet photosensitizers to generate singlet oxygen contain one and two light-harvesting antennas as well as

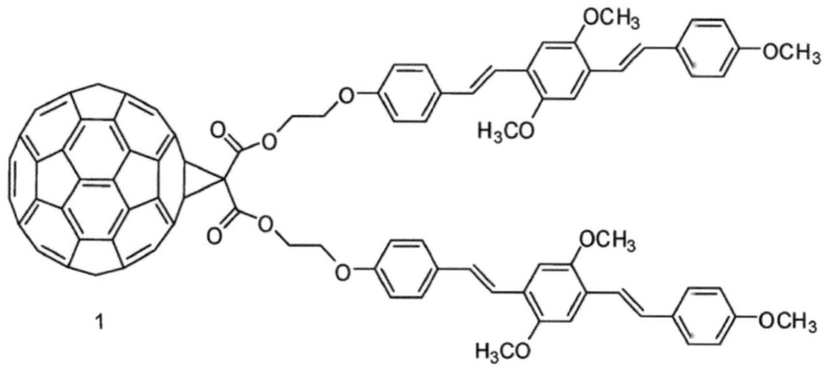

Figure 5. Fullerene-distyrylbenzene conjugate [105] (Reprinted by permission of RSC).

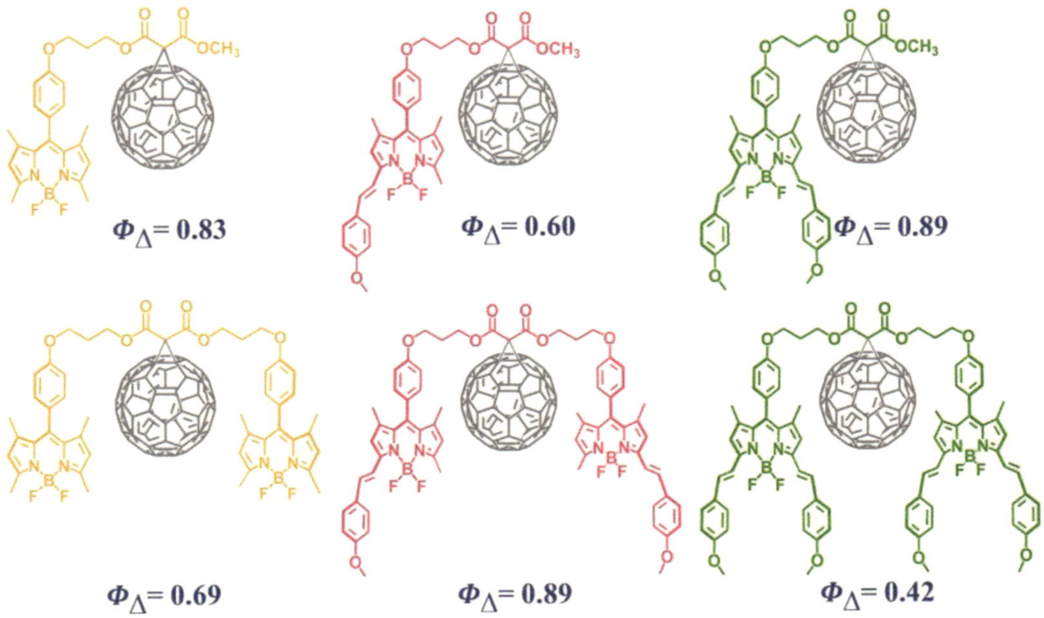

Figure 6. Heavy atom-free BODIPY-fullerene triplet photosensitizers [25, 26].

associated with different absorption wavelengths were successfully designed and synthesized in our laboratory (**Figure 6**).

Acknowledgements

The author would like to thank to Gebze Technical University.

Conflict of interest

There is no conflict of interest.

Author details

Elif Okutan

Address all correspondence to: eokutan@gtu.edu.tr

Gebze Technical University, Faculty of Science, Department of Chemistry Gebze, Kocaeli, Turkey

References

[1] Kroto HW, Heath JR, O'Brien SC, Curl RF, Smalley RE. C60: Buckminsterfullerene. Nature. 1985;**318**:162-163. DOI: 10.1038/318162a0

[2] Prato M. [60] fullerene chemistry for materials science applications. Journal of Materials Chemistry. 1997;**7**:1097-1109. DOI: 10.1039/A700080D

[3] Da Ros T, Prato M. Medicinal chemistry with fullerenes and fullerene derivatives. Chemical Communications. 1999:663-669. DOI: 10.1039/A809495K

[4] Fagan PJ, Calabrese JC, Malone B. Metal complexes of buckminsterfullerene (C60). Accounts of Chemical Research. 1992;**25**:134-142. DOI: 10.1021/ar00015a006

[5] Hirsch A, editor. Fullerenes and Related Structures. Topics in Current Chemistry. 1st ed. Berlin: Springer; 1999. DOI: 10.1007/3-540-68117-5

[6] Taylor R. Lecture Notes on Fullerene Chemistry. 1st ed. London: Imperial College Press; 1999

[7] Diederich F, Thilgen C. Covalent fullerene chemistry. Science. 1996;**271**:317-323. DOI: 10.1126/science.271.5247.317

[8] Lopez AM, Mateo-Alonso A, Prato M. Materials chemistry of fullerene C_{60} derivatives. Journal of Materials Chemistry. 2011;**21**:1305-1318. DOI: 10.1039/C0JM02386H

[9] Yadav BC, Kumar R. Structure, properties and applications of fullerenes. International Journal of Nanotechnology and Applications. 2008;**2**:15-24

[10] Foote CS. Photophysical and Photochemical Properties of Fullerenes. Topics in Current Chemistry. Vol. 169. Berlin, Heidelberg: Springer; 1994. pp. 347-363. DOI: 10.1007/3-540-57565-0_80

[11] Sun YP. Molecular and supramolecular photochemistry. In: Organic Photo Chemistry. New York: CRC Press; 1997. pp. 325-390

[12] Palit DK, Sapre AV, Mittal JP, Rao CNR. Photophysical properties of the fullerenes, C60 and C70. Chemical Physics Letters. 1992;**195**:1-6. DOI: 10.1016/0009-2614(92)85900-U

[13] Ajie H, Alvarez MM, Anz SJ, Beck RD, Diederich F, Fostiropoulos Huffman KDR, Krätschmer W, Rubin Y, Shriver KE, Sensharma D, Whetten RL. The Journal of Physical Chemistry. 1990;**94**:8630-8633. DOI: 10.1021/j100387a004

[14] Imahori H, Sakata Y. Donor-linked fullerenes: Photoinduced electron transfer and its potential application. Advanced Materials. 1997;**9**:537-546. DOI: 10.1002/adma.1997009 0704

[15] Imahori H, Sakata Y. Fullerenes as novel acceptors in photosynthetic electron transfer. European Journal of Organic Chemistry. 1999;**10**:2445-2457. DOI: 10.1002/(SICI)1099-0690(199910)1999:10<2445::AID-EJOC2445>3.0.CO;2-G

[16] Guldi DM. Fullerene–porphyrin architectures; photosynthetic antenna and reaction center models. Chemical Society Reviews. 2002;**31**:22-36. DOI: 10.1039/B106962B

[17] Guldi DM, Kamat PV. Fullerenes, Chemistry, Physics and Technology. New York: Wiley; 2000. 225p. DOI: 10.1002/aoc.173

[18] Medvedev ES, Osherov VI. Radiationless Transition in Polyatomic Molecules. Moscow: Springer-Verlag Berlin Heidelberg; 1983

[19] Couris S, Koudoumas E, Ruth AA, Leach S. Concentration and wavelength dependence of the effective third-order susceptibility and optical limiting of C_{60} in toluene solution. Journal of Physics B: Atomic, Molecular and Optical Physics. 1995;**8**:4537-4554. DOI: 10.1088/0953-4075/28/20/015

[20] Sun YP, Riggs JE, Liu B. Optical limiting properties of [60]fullerene derivatives. Chemistry of Materials. 1997;**9**:1268-1272. DOI: 10.1021/cm960650v

[21] Wray JE, Liu KC, Chen CH, Garret WR, Payne MG, Goedert R, Templeton D. Optical power limiting of fullerenes. Applied Physics Letters. 1994;**64**:2785-2787. DOI: 10.1063/1.111470

[22] Arbogast JW, Darmanyan AP, Foote CS, Diederich FN, Whetten RL, Rubin Y, Alvarez MM, Anz SJ. Photophysical properties of sixty atom carbon molecule (C60). Journal of Physical Chemistry. 1991;**95**:11-12. DOI: 10.1021/j100154a006.

[23] Arbogast JW, Foote CS. Photophysical properties of C70. Journal of American chemical Society. 1991;**113**:8886-8889. DOI: 10.1021/ja00023a041

[24] Tanigaki K, Ebbesen TW, Kuroshima S. Picosecond and nanosecond studies of the excited state properties of C70. Chemical Physics Letters. 1991;**185**:189-192. DOI: 10.1016/S0009-2614(91)85045-X

[25] Ünlü H, Okutan E. Novel distyryl BODIPY–fullerene dyads: Preparation, characterization and photosensitized singlet oxygen generation efficiency. New Journal of Chemistry. 2017;**41**:10424-10431. DOI: 10.1039/C7NJ02010D

[26] Ünlü H, Okutan E. Preparation of BODIPY- fullerene and monostyryl BODIPY-fullerene dyads as heavy atom free singlet oxygen generators. Dyes and Pigments. 2017;**142**:340-349. DOI: 10.1016/j.dyepig.2017.03.055

[27] Wang Y. Photoconductivity of fullerene-doped polymers. Nature. 1992;**356**:585-587. DOI: 10.1038/356585a0

[28] Smilowitz L, Sarıçiftçi NS, Wu R, Gettinger C, Heeger AJ, Wudl F. Photoexcitation spectroscopy of conducting-polymer-C60 composites: Photoinduced electron transfer. Physical Review B: Condensed Matter and Materials Physics. 1993;**47**:13835-13842. DOI: 10.1103/PhysRevB.47.13835

[29] Komamine S, Fujitsuka M, Ito O, Itaya A. Photoinduced electron transfer between C_{60} and carbazole dimer compounds in a polar solvent. Journal of Photochemistry and Photobiology A: Chemistry. 2000;**135**:111-117. DOI: 10.1016/S1010-6030(00)00290-2

[30] Singh Rachford TN, Castellano FN. Photon upconversion based on sensitized triplet-triplet annihilation. Coordination Chemistry Reviews. 2010;**254**:2560-2573. DOI: 10.1016/j.ccr.2010.01.003

[31] Zhao J, Ji S, Guo H. Triplet-triplet annihilation based upconversion: From triplet sensitizers and triplet acceptors to upconversion quantum yields. RSC Advanced. 2011;**1**:937-950. DOI: 10.1039/C1RA00469G

[32] Khnayzer RS, Blumhoff J, Harrington JA, Haefele A, Deng F, Castellano FN. Upconversion-powered photoelectrochemistry. Chemical Communications. 2012;**48**:209-211. DOI: 10.1039/C1CC16015J

[33] Ceroni P. Energy up-conversion by low-power excitation: New applications of an old concept. Chemistry A European Journal. 2011;**17**:9560-9564. DOI: 10.1002/chem.20110 1102

[34] Baluschev S, Miteva T, Yakutkin V, Nelles G, Yasuda A, Wegner G. Up-conversion fluorescence: Noncoherent excitation by sunlight. Physical Review Letters. 2006;**97**: 143903-143905. DOI: 10.1103/PhysRevLett.97.143903

[35] Chen HC, Hung CY, Wang KH, Chen HL, Fann WS, Chien FC, Chen P, Chow TJ, Hsu CP, Sun SS. White light emission from an upconverted emission with an organic triplet sensitizer. Chemical Communications. 2009:4064-4066. DOI: 10.1039/B905572J

[36] Wu W, Zhao J, Sun J, Guo S. Light-harvesting fullerene dyads as organic triplet photosensitizers for triplet-triplet annihilation upconversions. The Journal of Organic Chemistry. 2012;**77**:5305-5312

[37] Fraelich MR, Weisman RB. Triplet states of fullerene C60 and C70 in solution: Long intrinsic lifetimes and energy pooling. Journal of Physical Chemistry. 1993;**97**:11145-11147. DOI: 10.1021/j100145a002

[38] Ebbesen TW, Tanigaki K, Kuroshima S. Excited-state properties of C_{60}. Chemical Physics Letters. 1991;**181**:501-504. DOI: 10.1016/0009-2614(91)80302-E

[39] Biczok L, Linschitz H, Walter RI. Extinction coefficients of C_{60} triplet and anion radical, and one-electron reduction of the triplet by aromatic donors. Chemical Physics Letters. 1992;**195**:339-346. DOI: 10.1016/0009-2614(92)85613-F

[40] Kajii Y, Nakagawa T, Suzuki S, Achiba Y, Obi K, Shibuya K. Transient absorption, lifetime and relaxation of C_{60} in the triplet state. Chemical Physics Letters. 1991;**181**: 100-104. DOI: 10.1016/0009-2614(91)90339-B

[41] Terazima M, Hirota N, Shinohara H, Saito Y. Photothermal investigation of the triplet state of carbon molecule (C60). The Journal of Physical Chemistry. 1991;**95**:9080-9085. DOI: 10.1021/j100176a013

[42] Closs GL, Gautam P, Zhang D, Krusic PJ, Hill SA, Wasserman E. Steady-state and time-resolved direct detection EPR spectra of fullerene triplets in liquid solution and glassy matrixes: Evidence for a dynamic Jahn-teller effect in triplet C60. The Journal of Physical Chemistry. 1992;**96**:5228-5231. DOI: 10.1021/j100192a011

[43] Dimitrijevic NM, Kamat PV. Triplet excited state behavior of fullerenes: Pulse radiolysis and laser flash photolysis of fullerenes (C60 and C70) in benzene. The Journal of Physical Chemistry. 1992;**96**:4811-4814. DOI: 10.1021/j100191a017

[44] Palit D, Sapre AV, Mittal JP, Rao CNR. Photophysical properties of the fullerenes, C_{60} and C_{70}. Chemical Physics Letters. 1992;**195**:1-6. DOI: 10.1016/0009-2614(92)85900-U

[45] Gijzeman OLJ, Kaufman F, Porter G. Oxygen quenching of aromatic triplet states in solution. Part 1. Journal of the Chemical Society, Faraday Transactions 2: Molecular and Chemical Physics. 1973;**69**:708-720. DOI: 10.1039/F29736900708

[46] Turro NJ. Modern Molecular Photochemistry. Menlopark CA: The Benjamin/Cummings Publishing Company; 1978

[47] Sandros K. Transfer of triplet state energy in fluid solutions. III. Reversible energy transfer. Acta Chemica Scandinavica. 1964;**18**:2355-2374. DOI: 10.3891/acta.chem.scand.18-2355

[48] Liddell PA, Sumida JP, Macpherson AN, Noss L, Seely GR, Clark KN, Moore AL, Moore TA, Gust D. Preparation and photophysical studies of porphyrin-C60 dyads. Photochemistry and Photobiology. 1994;**60**:537-541. DOI: 10.1111/j.1751-1097.1994.tb05145.x

[49] Kuciauskas D, Lin S, Seety GR, Moore AL, Moore TA, Gust D, Drovetskaya T, Reed CA, Boyd PDW. Energy and Photoinduced electron transfer in Porphyrin-fullerene dyads. Journal of Physical Chemistry. 1996;**100**:15926-15932. DOI: 10.1021/jp9612745

[50] Kafafi ZH, Lindle JR, Pong RGS, Bartoli FJ, Lingg LJ, Milliken J. Off-resonant nonlinear optical properties of C_{60} studied by degenerate four-wave mixing. 1992;**188**:492-496. DOI: 10.1016/0009-2614(92)80854-5

[51] Kajzar F, Taliani C, Zamboni R, Rossini S, Danieli R. Nonlinear optical properties of fullerenes. Synthetic Metals. 1996;**77**:257-263

[52] Wang Y, Cheng LT. Nonlinear optical properties of fullerenes and charge-transfer complexes of fullerenes. The Journal of Physical Chemistry. 1992;**96**:1530-1532

[53] Shulev VA, Filippov AK, Kamanina NV. Laser-induced processes in the IR range in Nanocomposites with fullerenes and carbon nanotubes. Technical Physics Letters. 2006; **32**:694-697. DOI: 10.1134/S1063785006080177

[54] Loboda O, Zalesny R, Avramopoulos A, Luis JM, Kirtman B, Tagmatarchis N, Reis H, Papadopoulos MG. Linear and nonlinear optical properties of [60]fullerene derivatives. The Journal of Physical Chemistry. A. 2009;**113**:1159-1170. DOI: 10.1021/jp808234x

[55] Kamanina N. Mechanisms of optical limiting in π conjugated organic system: Fullerene doped polyimide. Synthetic Metals. 2002;**127**:121-128. DOI: 10.1016/S0379-6779(01) 00598-7

[56] Kamanina NV. Properties of optical limitation of radiation in π-conjugated organic systems doped with fullerenes (the COANP-C70 system). Optics and Spektroscopy. 2001;**90**:931-937. DOI: 10.1134/1.1380795

[57] Kamanina NV. Nonlinear optical coefficients of polyimide doped with fullerenes. Synthetic Metals. 2003;**139**:547-550. DOI: 10.1016/S0379-6779(03)00358-8

[58] van Hal PA, Janssen RA, Lanzani G, Cerullo G, Zavelani-Rossi M, De Silvestri S. Two-step mechanism for the photoinduced intramolecular electron transfer in oligo (p-phenylene vinylene)-fullerene dyads. Physical Review B. 2001;**64**:075206-1-075206-7. DOI: 10.1103/PhysRevB.64.075206

[59] Imahori H, Hagiwara K, Akiyama T, Taniguchi S, Okada T, Sakata Y. Synthesis and Photophysical property of Porphyrin-linked fullerene. Chemistry Letters. 1995;**24**:265-266. DOI: 10.1246/cl.1995.265

[60] Linssen TG, Dürr K, Hanack M, Hirsch A. A green fullerene: Synthesis and electrochemistry of a Diels-Alder adduct of [60]fullerene with a Phthalocyanine. Journal of the Chemical Society, Chemical Communications. 1995;**0**:103-104. DOI: 10.1039/C39950000103

[61] Maggini M, Dono A, Scorrano G, Prato M. Synthesis of a [60]fullerene derivative covalently linked to a ruthenium(II) tris(bipyridine) complex. Journal of the Chemical Society, Chemical Communications. 1995;**0**:845-846. DOI: 10.1039/C39950000845

[62] Maggini M, Karlsson A, Scorrano G, Sandona G, Farina G, Prato M. Ferrocenyl fulleropyrrolidines: A cyclic voltammetry study. Journal of the Chemical Society, Chemical Communications. 1994;**0**:589-590. DOI: 10.1039/C39940000589

[63] Drovetskaya T, Reed CA, Boyd PDW. A fullerene Porphyrin conjugate. Tetrahedron Letters. 1995;**36**:7971-7974. DOI: 10.1016/0040-4039(95)01719-X

[64] Imahori H, Cardoso S, Tatman D, Lin S, Macpherson AN, Noss L, Seely GR, Sereno L, Chessa de Silber J, Moore TA, Moore AL, Gust D. Photoinduced electron transfer in a carotenobuckminsterfullerene dyad. Photochemistry and Photobiology. 1995;**62**:1009-1014. DOI: 10.1111/j.1751-1097.1995.tb02401.x

[65] Williams RM, Zwier JM, Verhoeven JW. Photoinduced Intramolecular electron transfer in a bridged C60 (acceptor)-aniline (donor) system; Photophysical properties of the first "active" fullerene Diad. Journal of the American Chemical Society. 1995;**117**:4093-4099. DOI: 10.1021/ja00119a025

[66] Imahori H, Sakata Y. Synthesis of closely spaced Porphyrin-fullerene. Chemistry Letters. 1996;**25**:199-200. DOI: 10.1246/cl.1996.199

[67] Williams RM, Koeberg M, Lawson JM, An YZ, Rubin Y, Paddon-Row MN, Verhoeven JW. Photoinduced electron transfer to C60 across extended 3- and 11-bond hydrocarbon bridges: Creation of a long-lived charge-separated state. Journal of Organic Chemistry. 1996;**61**:5055-5062. DOI: 0.1021/jo960678q

[68] Ziessel R, Allen BD, Rewinska DB, Harriman A. Selective triplet-state formation during charge recombination in a fullerene/Bodipy molecular dyad (Bodipy=Borondipyrromethene). Chemistry – A European Journal. 2009;**15**:7382-7393. DOI: 10.1002/chem.200900440

[69] van Haver P, van der Auweraer M, Viaene L, De Schryver FC, Verhoeven JW, Van Ramesdonk HJ. Enhanced intersystem crossing in 3-(1-pyrenyl)propylbromide. Chemical Physics Letters. 1992;**198**:361-366. DOI: 10.1016/0009-2614(92)85065-I

[70] Bartczak WM, Hummel A. Formation of singlet and triplet excited states on charge recombination in tracks of high-energy electrons in nonpolar liquids. A computer simulation study. Chemical Physics Letters. 1993;**208**:232-236. DOI: 10.1016/0009-2614(93) 89067-R

[71] Murakami M, Ohkubo K, Mandal P, Ganguly T, Fukuzumi S. Does bimolecular charge recombination in highly exergonic electron transfer afford the triplet excited state or the ground state of a photosensitizer? The Journal of Physical Chemistry A. 2008;**112**:635-642. DOI: 10.1021/jp0767718

[72] Lavrik NL. Analysis of the recombination probability of geminate triplet radical ion pairs at single encounter by means of magnetic field effects. The Journal of Chemical Physics. 2001;**114**:9492-9495. DOI: 10.1063/1.1365083

[73] Tero-Kubota S. Singlet and triplet energy splitting in the radical ion pairs generated by photoinduced electron-transfer reactions. Pure and Applied Chemistry. 2001;**73**:519-523. DOI: 10.1351/pac200173030519

[74] Nishimura T, Nakashima N, Mataga N. Intersystem crossing in anthracene-N,N-diethylaniline exciplex. Chemical Physics Letters. 1977;**46**:334-338. DOI: 10.1016/0009-2614(77)85275-5

[75] Mataga N, Migita M, Nishimura T. Picosecond chemistry of some exciplex systems. Journal of Molecular Structure. 1978;**47**:199-219. DOI: 10.1016/0022-2860(78)87185-3

[76] Mataga N. Picosecond laser photolysis studies upon charge separation and intersystem crossing processes in some exciplex systems. Radiation Physics and Chemistry. 1977;**21**: 83-89. DOI: 10.1016/0146-5724(83)90133-4

[77] Swinnen AM, Van der Auweraer M, De Schryver FC, Nakatani K, Okada T, Mataga N. Photophysics of the intramolecular exciplex formation in .Omega.-(1-pyrenyl)-.Alpha.-(dimethylamino)alkanes. Journal of the American Chemical Society. 1987;**109**:321-330. DOI: 10.1021/ja00236a005

[78] Kuciauskas D, Liddell PA, Moore AL, Moore TA, Gust D. Magnetic switching of charge separation lifetimes in artificial photosynthetic reaction Centers. Journal of the American Chemical Society. 1998;**120**:10880-10886. DOI: 10.1021/ja981848e

[79] Kobori Y, Yamauchi S, Akiyama K, Tero-Kubota S, Imahori H, Fukuzumi S, Norris JR. Primary charge-recombination in an artificial photosynthetic reaction center. PNAS. 2005;**102**:10017-10022. DOI: 10.1073.pnas.0504598102

[80] Fukuzumi S, Guldi DM. In: Balzani V, editor. Electron Transfer in Chemistry. Vol. 2. Weinheim: Wiley-VCH; 2001. pp. 270-337

[81] Umeyama T, Imahori H. Carbon nanotube-modified electrodes for solar energy conversion. Energy & Environmental Science. 2008;**1**:120-133. DOI: 10.1039/B805419N

[82] Guldi DM. Fullerene–porphyrin architectures; photosynthetic antenna and reaction center models. Chemical Socety Review. 2002;**31**:22-36. DOI: 10.1039/B106962B

[83] Chitta R, D'Souza F. Self-assembled tetrapyrrole–fullerene and tetrapyrrole–carbon nanotube donor-acceptor hybrids for light induced electron transfer applications. Journal of Materials Chemistry. 2008;**18**:1440-1471. DOI: 10.1039/B717502G

[84] Maligaspe E, Kumpulainen T, Subbaiyan NK, Zandler ME, Lemmetyinen H, Tkachenko NV, D'Souza F. Electronic energy harvesting multi BODIPY-zinc porphyrin dyads accommodating fullerene as photosynthetic composite of antenna-reaction center. Physical Chemistry Chemical Physics. 2010;**12**:7434-7444. DOI: 10.1039/C002757J

[85] Rio Y, Seitz W, Gouloumis A, Vazquez P, Sessler JL, Guldi DM, Torres T. Panchromatic Supramolecular fullerene-based donor–acceptor assembly derived from a peripherally substituted Bodipy–zinc Phthalocyanine dyad. Chemistry - A European Journal. 2010; **16**:1929-1940. DOI: 10.1002/chem.200902507

[86] Liu JY, El-Khouly ME, Fukuzumi S, Ng DKP. Photoinduced electron transfer in a Distyryl BODIPY–fullerene dyad. Chemistry, an Asian Journal. 2011;**6**:174-179. DOI: 10.1002/asia.201000537

[87] Huang D, Zhao J, Wu W, Yi X, Yang P, Ma J. Visible-light-harvesting Triphenylamine Ethynyl C60-BODIPY dyads as heavy-atom-free organic triplet photosensitizers for triplet-triplet annihilation Upconversion. Asian Journal of Organic Chemistry. 2012;**1**: 264-273. DOI: 10.1002/ajoc.201200062

[88] Chen CY, Chen JG, Wu SJ, Li JY, Wu CG, Ho KC. Multifunctionalized ruthenium-based super sensitizers for highly efficient dye-sensitized solar cells. Angewanthe Chemie. 2008;**120**:7452-7455. DOI: 10.1002/ange.200802120

[89] Monguzzi A, Mezyk J, Scotognella F, Tubino R, Meinardi F. Upconversion-induced fluorescence in multicomponent systems: Steady-state excitation power threshold. Physical Review B. 2008;**78**:195112. DOI: 10.1103/PhysRevB.78.195112

[90] Wu W, Zhao J, Guo H, Sun J, Ji S, Wang Z. Long-lived room-temperature near-IR phosphorescence of BODIPY in a visible-light-harvesting N^C^N PtII–Acetylide complex with a directly Metalated BODIPY Chromophore. Chemistry—a European Journal. 2012;**18**:1961-1968. DOI: 10.1002/chem.201102634

[91] Singh-Rachford TN, Castellan FN. Photon upconversion based on sensitized triplet–triplet annihilation. Coordination Chemistry Reviews. 2010;**254**:2560-2573. DOI: 10.1016/j.ccr.2010.01.003

[92] Ji S, Wu W, Wu W, Guo H, Zhao J. Ruthenium(II) Polyimine Complexes with a Long-Lived 3IL Excited State or a 3MLCT/3IL Equilibrium: Efficient Triplet Sensitizers for

Low-Power Upconversion. Vol. 50. Angewandte Chemie International Edition; 2011. pp. 1626-1629. DOI: 10.1002/anie.201006192

[93] Chen HC, Hung CY, Wang KH, Chen HL, Fann WS, Chien FC, Chen P, Chow TJ, Hsu CP, Sun SS. White-light emission from an upconverted emission with an organic triplet sensitizer. Chemical Communications. 2009;**0**:4064-4066. DOI: 10.1039/B905572J

[94] Yang P, Wu W, Zhao J, Huang D, Yi X. Using C60-bodipy dyads that show strong absorption of visible light and long-lived triplet excited states as organic triplet photosensitizers for triplet–triplet annihilation upconversion. Jounal of Materials Chemistry. 2012;**22**:20273-20283. DOI: 10.1039/c2jm34353c

[95] Moor K, Kim JH, Snow S, Kim JH. [C70] Fullerene-sensitized triplet–triplet annihilation upconversion. Chemical Communications. 2013;**49**:10829-10831. DOI: 10.1039/c3cc46598e

[96] Guo S, Sun J, Ma L, You W, Yang P, Zhao J. Visible light-harvesting naphthalenediimide (NDI)-C60 dyads as heavy-atom-free organic triplet photosensitizers for triplet-triplet annihilation based upconversion. Dyes and Pigments. 2013;**96**:449-458. DOI: 10.1016/j.dyepig.2012.09.008

[97] Lim GN, Maligaspe E, Zandler ME, D'Souza F. A Supramolecular tetrad featuring covalently linked Ferrocene–zinc Porphyrin–BODIPY coordinated to fullerene: A charge stabilizing, photosynthetic antenna–reaction Center mimic. Chemistry – A European Journal. 2014;**20**:17089-17099. DOI: 10.1002/chem.201404671

[98] Sera N, Tokiwa H, Miyata N. Mutagenicity of the fullerene C_{60}-generated singlet oxygen dependent formation of lipid peroxides. Carrinogenesis. 1996;**17**:2163-2169. DOI: 10.1093/carcin/17.10.2163

[99] Hamano T, Okuda K, Mashino T, Hirobe M, Arakane K, Ryu A, Mashikoc S, Nagano T. Singlet oxygen production from fullerene derivatives: Effect of sequential functionalization of the fullerene core. Chemical Communications. 1997;**0**:21-22. DOI: 10.1039/A606335G

[100] Prat F, Stackow R, Bernstein R, Qian W, Rubin Y, Foote CS. Triplet-state properties and singlet oxygen generation in a homologous series of functionalized fullerene derivatives. Physical Chemistry A. 1999;**103**:7230-7235. DOI: 10.1021/jp991237o

[101] Prat F, Martí C, Nonell S, Zhang X, Foote CS, Moreno RG, Bourdelande JL, Font J. C_{60} fullerene-based materials as singlet oxygen $O_2(^1\Delta_g)$ photosensitizers: A time-resolved near-IR luminescence and optoacoustic study. Physical Chemistry Chemical Physics. 2001;**3**:1638-1643. DOI: 10.1039/B009484F

[102] Bourdelande JL, Font J, Gonzalez-Moreno R. Fullerene C-60 bound to insoluble hydrophilic polymer: Synthesis, photophysical behavior, and generation of singlet oxygen in water suspensions. Helvetica Chimica Acta. 2001;**84**:3488-3494. DOI: 10.1002/1522-2675(20011114)84:11<3488::AID-HLCA3488>3.0.CO;2-7

[103] Murata Y, Komatsu K. Photochemical reaction of the open-cage fullerene derivative with singlet oxygen. Chemistry Letters. 2001;**9**:896-897. DOI: 10.1246/cl.2001.896

[104] McCluskey DM, Smith TN, Madasu PK, Coumbe CE, Mackey MA, Fulmer PA, Wynne JH, Stevenson S, Phillips JP. Evidence for singlet-oxygen generation and Biocidal activity in Photoresponsive metallic nitride fullerene-polymer adhesive films. ACS Applied Materials & Interfaces. 2009;**1**:882-887. DOI: 10.1021/am900008v

[105] Collini E, Fortunati I, Scolaro S, Signorini R, Ferrante C, Bozio R, Fabbrini G, Maggini M, Rossi E, Silvestrini S. A fullerene-distyrylbenzene photosensitizer for two-photon promoted singlet oxygen production. Physical Chemistry Chemical Physics. 2010;**12**:4656-4666. DOI: 10.1039/b922740g

[106] Gobeze HB, Tram T, Chandra BKC, Cantu RR, Karr PA, D'Souza F. Singlet oxygen generation and Photoinduced charge separation of tetra Polyethyleneglycol functionalized zinc Phthalocyanine-fullerene dyad. Chinese Journal of Chemistry. 2016;**34**:969-974. DOI: 10.1002/cjoc.201600403

[107] Montagnon T, Tofi M, Vassilikogiannakis G. Using singlet oxygen to synthesize Polyoxygenated natural products from furans. Accounts on Chemical Research. 2008; **41**:1001-1011. DOI: 10.1021/ar800023v

[108] Kyriakopoulos J, Papastavrou AT, Panagiotoua GD, Tzirakis MD, Triantafyllidis K, Alberti MN, Bourikas K, Kordulis C, Orfanopoulos M, Lycourghiotis A. Deposition of fullerene C60 on the surface of MCM-41 via the one-step wet impregnation method: Active catalysts for the singlet oxygen mediated photooxidation of alkenes. Journal of Molecular Catalysis A: Chemical. 2014;**381**:9-15. DOI: 10.1016/j.molcata.2013.09.036

[109] Huang L, Cui X, Therrien B, Zhao J. Energy-Funneling-based broadband visible-light-absorbing Bodipy–C_{60} triads and tetrads as dual functional heavy-atom-free organic triplet photosensitizers for Photocatalytic organic reactions. Chemistry – A European Journal. 2013;**19**:17472-17482. DOI: 10.1002/chem.201302492